RETHINKING FAECAL S␣␣␣␣␣␣␣␣␣␣␣␣ENT
IN EMERGENCY SETTINGS
Decision Support Tools and Smart Technology Applications
for Emergency Sanitation

Fiona ZAKARIA

RETHINKING FAECAL SLUDGE MANAGEMENT IN EMERGENCY SETTINGS

Decision Support Tools and Smart Technology Applications
for Emergency Sanitation

DISSERTATION

Submitted in fulfilment of the requirements of
the Board for Doctorates of Delft University of Technology
and
of the Academic Board of the IHE Delft
Institute for Water Education
for
the Degree of DOCTOR
to be defended in public on
Friday, 28 June, at 12.30 hours
in Delft, the Netherlands

by

Fiona ZAKARIA

Master of Science in Water Science, Policy and Management, University of Oxford, UK
Master of Science in Sanitary and Environmental Engineering, University Putra Malaysia
born in Bandung, Indonesia

This dissertation has been approved by the
promotor: Prof. dr. D. Brdjanovic

Composition of the doctoral committee:

Rector Magnificus TU Delft   Chairman
Rector IHE Delft   Vice-Chairman
Prof.dr. D. Brdjanovic   IHE Delft/ TU Delft, promotor

Independent members:
Prof.dr.ir. J.B. van Lier   TU Delft
Prof.dr.ir. G.J.F.R. Alaerts   IHE Delft / TU Delft
Prof.dr. M. von Sperling   Federal University of Mina Gerais, Brasil
Prof.dr.ir. G. Zeeman   Wageningen University & Research
Prof.dr.ir. M.K. de Kreuk   TU Delft, reserve member

This research was conducted under the auspices of the SENSE Research School for
Socio-Economic and Natural Sciences of the Environment

CRC Press/Balkema is an imprint of the Taylor & Francis Group, an informa business

Published by:
CRC Press/Balkema
Schipholweg 107C, 2316 XC, Leiden, the Netherlands
Pub.NL@taylorandfrancis.com
www.crcpress.com – www.taylorandfrancis.com
ISBN 978-0-367-36181-5

the water and sanitation crisis claims more lives through diseases than any war claims through weapons (UNHDR's Water Fact 2012)

Table of content

Thesis summary

The development of technology in the emergency sanitation sector has not been emphasised sufficiently considering that the management of human excreta is a basic requirement for every person. The lack of technology tailored to emergency situations complicates efforts to cater for sanitation needs in challenging humanitarian crisis. Sanitation response together with the provision of clean water and hygiene promotion are considered life-saving efforts in emergencies. Nevertheless, in an emergency, there is regularly lack of means and limited planning time available to provide an effective and safe sanitation response.

Reviewing the existing practices, the emergency toilet options consist of very basic provisions, primarily trench and pit latrines. Whenever it is not possible to dig a pit or trench, the option left is using container based sanitation. This type of sanitation in particular requires a collection or emptying plan, and a subsequent treatment and safe disposal plan, which is usually overlooked in the realm of an emergency where there is limited time to plan for any requirements after toilet provisions.

With the above-mentioned concerns in mind, this study focused on the development of a smart emergency toilet termed the eSOS (emergency sanitation operation system) smart toilet to address the limitation in technical options. This toilet is based on the eSOS concept that takes into account the entire sanitation chain, which is the required processing of human excreta from toilet until safe disposal (downstream process). The initial design was the basis to the experimental toilet prototype, which was then tested and evaluated under real use conditions in an emergency camp in Tacloban City, in the Philippines.

This field research in the Philippines evaluated different design related aspects of the toilet, such as the operation, user acceptance, and specific smart features. The aim was also to obtain new knowledge using the toilet's sophisticated monitoring system, such as information on the toilet use and the characterisation of generated faecal sludge and urine streams under real emergency conditions.

This PhD study also addresses the limited time for planning in emergencies by developing a decision support system (DSS) to help quick selection of optimal sanitation options. The aim was to enable users of the DSS to plan their emergency sanitation response within the shortest time possible. The DSS tool gathered all the technical options suitable for use in emergencies and organised them into corresponding function groups in the sanitation chain. The user is asked for input to relate the suitability of the technical options with the user's scenario. The tool subsequently guides the user to plan each chain element to make up planning for the complete chain. The user can compare workable sanitation alternatives presented in a few chains. The selection process is an iterative process, by giving an overview of the impact of the different sanitation technical options in their planned sanitation process. The tool outcome is several evaluated sanitation chains. The evaluation uses a rating system, which is also completed by the users. The outcome gives the highest rated sanitation system as the most suitable sanitation technical option in the users' given scenario.

The sophistication of the eSOS monitoring system that can measure and track the material flow gave the opportunity to estimate costs from all activities in one functioning sanitation chain. Cost components are regularly missing in general sanitation planning. Often, the cost estimation is only provided for a single sanitation chain, instead of costs for the entire sanitation chain. A financial flow simulator called eSOS Monitor was developed to address this gap in sanitation chain cost estimation. Additionally, eSOS Monitor adopted sanitation technology selection by means of the previously developed DSS and subsequently calculates the costs for each chain. The cost summary then also calculates several financial indicators such as the breakeven time and returns that are useful for parties interested in investing in the business.

The study aims to contribute toward a better emergency sanitation response by application of technology advances. The eSOS Smart Toilet offers a toilet with monitoring system that ensures 'just-in-time' or responsive maintenance, amongst other smart features. Such system would ensure optimum toilet usage whilst maintaining sanitary condition despite high number of toilet visits. Efforts were made to extend the results application to benefit situations beyond emergency, by expanding eSOS Monitor as a financial flow simulator.

There is a large innovation gap in emergency sanitation, as well as innovation gaps in sanitation in general. Despite recently initiated efforts such as 'Re-invent Toilet Challenge' by Bill & Melinda Gates Foundation, agenda shifts from water to sanitation, inclusion of sanitation in Milennium Development Goals (MDG) 2015 and then Sustainable Development Goals (SDG) 2030 and many other initiatives, there remains gaps to cover, rooms for improvements, and progress to be made. This research learned that cooperation and coordinated efforts to be amongst the key factor in realizing successful innovations.

Samenvatting

De technologische ontwikkeling die nodig is om op een hygiënische en efficiënte manier om te gaan met menselijke uitwerpselen, een basisvoorziening voor iedereen, krijgt onvoldoende aandacht in de sanitaire sector. Vooral de inspanningen om tegemoet te komen aan de sanitaire behoeften in noodsituaties worden bemoeilijkt door het gebrek aan passende technologie. Het aanbieden van sanitaire voorzieningen, in combinatie met schoon drinkwater en het stimuleren van hygiënische omstandigheden, wordt in noodgevallen beschouwd als levensreddende inspanningen. Desalniettemin is er in een noodsituatie regelmatig gebrek aan sanitaire voorzieningen en is er beperkt tijd beschikbaar om dezen op een effectieve en veilige wijze in te plannen en te verstrekken.

Wanneer de huidige aanpak onder de loep wordt genomen, blijkt dat de gangbare opties voor noodtoiletten zeer basaal zijn, voornamelijk greppels of putten gegraven in de grond met een latrine erboven. Als het niet mogelijk is om een put of greppel te graven, wordt er gebruikt gemaakt van sanitaire voorzieningen gebaseerd op containers. Dit type sanitair vereist, nog meer dan latrines boven een put, een plan van aanpak om ze te legen, de inhoud te verzamelen en op een veilige manier af te voeren. In het geval van een noodsituatie wordt er tijdens en na het voorzien plaatsen van de toiletten te weinig rekening gehouden met het vereiste onderhoud.

Met de bovengenoemde zorgen in het achterhoofd, richtte deze studie zich op de ontwikkeling van een slim noodtoilet, genaamd het eSOS-toilet (emergency sanitation operation system) om de onderhoudsproblematiek op een technisch ondersteunde manier aan te pakken. Dit toilet is gebaseerd op het eSOS-concept, dat rekening houdt met de volledige sanitatieketen, van toilet tot en met de veilige verwijdering van menselijke uitwerpselen. Op de basis van het ontwerp van het eerste toilet werd een experimenteel prototype gemaakt, dat vervolgens werd getest en geëvalueerd onder reële gebruiksomstandigheden in een noodkamp in de stad Tacloban, op de Filippijnen.

Het veldonderzoek in de Filippijnen evalueerde verschillende ontwerp gerelateerde aspecten van het toilet, zoals de werking, gebruikersacceptatie, en specifieke slimme functies. Naast het testen van het ontwerp, was het tevens de bedoeling om nieuwe kennis op te doen met betrekking tot toiletgebruik in het algemeen en de karakterisering van gegenereerde fecale slib- en urinestromen onder reële noodomstandigheden, met behulp van het geavanceerde monitoringsysteem van het toilet.

Het promotieonderzoek richtte zich ook op het probleem van de beperkte tijd die beschikbaar is voor planning tijdens noodsituaties, door de ontwikkeling van een beslissingsondersteunend systeem (decision support system, DSS), om te helpen bij een snelle selectie van optimale sanitaire oplossingen. Het doel was om gebruikers van de DSS in staat te stellen om in korte tijd een overzicht te genereren van mogelijke toepasbare sanitaire voorzieningen in de context van noodhulp. De DSS-tool laat alle technische opties zien die geschikt zijn voor gebruik in noodsituaties en ordent ze in overeenkomstige functiegroepen in de sanitatieketen. De gebruiker wordt gevraagd om input te geven om de geschiktheid van de technische opties te relateren aan de randvoorwaarden bepaald door de specifieke situatie. De tool begeleidt vervolgens de gebruiker om elk ketenelement te plannen wat leidt tot de totale planning voor de volledige keten. De gebruiker kan werkbare sanitaire alternatieven vergelijken die voor de verschillende ketens worden gepresenteerd. Het selectieproces wat daarop volgt is iteratief, gebruik makende van de impact van de verschillende sanitaire technische opties. Het resultaat van de tool zijn verschillende geëvalueerde sanitaire voorzieningen. De evaluatie maakt gebruik van een beoordelingssysteem dat kan worden aangepast door de gebruikers. Het resultaat geeft het hoogst beoordeelde sanitaire systeem als de meest geschikte optie in het gegeven scenario van de gebruiker.

Het eSOS-controlesysteem, dat de materiaalstromen kan meten en volgen, bied de mogelijkheid om de kosten van alle activiteiten in één functionerende sanitatieketen in te schatten. Regelmatig blijkt dat de kostenraming alleen verstrekt wordt voor een enkele stap in de keten, in plaats van de alle kosten te berekenen. Een financiële flowsimulator, eSOS Monitor genaamd, werd ontwikkeld om deze discrepantie in de schatting van de kosten voor de sanitaire voorzieningen aan te pakken. De eSOS Monitor kan de door de DSS geselecteerde technologie gebruiken en berekent vervolgens de kosten voor elke stap in de keten. De kostensamenvatting laat vervolgens de verschillende financiële indicatoren zien, zoals het break-evenpoint en rendementen, die nuttig zijn voor partijen die geïnteresseerd zijn om te investeren in de voorzieningen.

Het promotie onderzoek had als doel om bij te dragen aan betere voorzieningen in het kader van noodhulp sanitatie, door toepassing van technologische ontwikkelingen. De eSOS Smart Toilet biedt een toilet met bewakingssysteem dat zorgt voor 'just-in-time' of responsief onderhoud, naast andere slimme functies. Een dergelijk systeem maakt een optimaal toiletgebruik mogelijk en zorgt ervoor dat hygiënische omstandigheden behouden blijven ondanks een hoog aantal en wisselende toiletbezoeken. Er is ook gekeken naar de mogelijkheid om de resultaten toe te passen in andere situaties waar geen sprake is van een noodsituatie, door eSOS Monitor uit te breiden tot een financiële flowsimulator.

Er is een grote innovatiekloof wat betreft sanitaire voorzieningen in het algemeen, en in het bijzonder tijdens noodsituaties. Ondanks recente internationale inspanningen, zoals 'Re-invent Toilet Challenge' door Bill en Melinda Gates Foundation, verschuiving van de aandacht van drinkwater naar sanitaire voorzieningen, opname van sanitaire voorzieningen in de ontwikkelingsdoelstellingen (MDG 2015, SDG 2030) en vele andere initiatieven, is er nog steeds ruimte voor verbetering, en zijn er nog vorderingen te maken. Dit onderzoek heeft aangetoond dat samenwerking en gecoördineerde inspanningen een van de belangrijkste factoren zijn bij het realiseren van succesvolle innovatie

1

General introduction

1.1 Emergencies following natural and anthropological disasters and displaced populations

The world has seen an escalating number of disasters over recent decades, from natural as well as anthropological origins. There have been disasters in South Asia (Tsunami, 2004) and in Haiti (Earthquake, 2010) that each caused hundreds of thousands of deaths, whilst floods and droughts have occurred more frequently. Anthropological disasters, for instance armed conflicts are affecting millions of people globally. Moreover, with the threat of unpredictable weather changes, global warming, continuing earth crust movement and political uncertainties, the number of disasters will likely increase. Figure 1-1 illustrates the number of natural disasters from 1980 to 2014, showing an increasing trend. The escalation of anthropological disasters via the trend of global displacements also indicates an increase (see Figure 1-3).

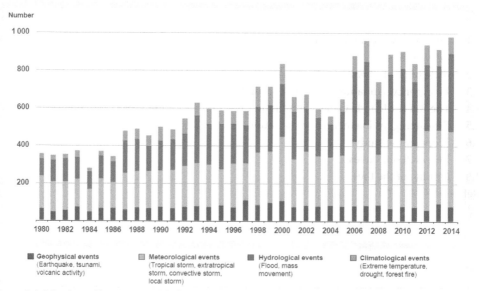

Figure 1-1: Number of loss events globally (1980–2014); Source: NatCatSERVICE (2015) – As at January 2015

The scale of a disaster may be measured by the death toll, economic loss and numbers of affected people. Analysing natural disasters in the last decade, World Disaster Report 2010 (IFRC 2010) established the following.

- Earthquakes killed the most people from 2000 to 2008 – an average of around 50,000 people a year.
- Floods, meanwhile, have affected the most significant number of people – an average of 99 million people a year.
- The costliest urban disaster of the last decade was the Bam earthquake in Iran, in 2003, which left damages totalling US$500 M.
- The deadliest disaster was the South Asian tsunami in 2004, which affected seven countries and killed 226,408 people.

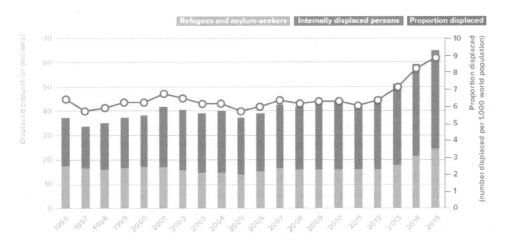

Figure 1-2 Trend of global displacement and proportion displaced 1996 – 2015 (UNHCR 2016)

Disasters cause people to flee from their homes to seek refuge in a safer place in or outside their country of origin. Displaced people within the boundary of their original country are referred to as internally displaced persons (IDP), while those displaced to another country are termed 'refugees.' Unlike for death toll, the number of displaced people has been more problematic to document as they change over time. By the end of 2014, the United Nations High Commissioner for Refugees (UNHCR) reported that the number of displacements worldwide was at all time high with 59.5 million, and was likely to deteriorate further (UNHCR 2015). The major contribution was from war events in Syria and several surrounding countries in the Middle East. The trend forecast was proven to be correct through their later report in 2016 (UNHCR 2016), as observed in Figure 1-2.

Complicated disasters, known as "complex emergencies" among humanitarian organisations, (Burkholder & Toole 1995) are attended by responses referred to as 'emergency responses'. Complex emergencies are defined as "relatively acute situations affecting large civilian populations, usually involving a combination of war or civil strife, food shortages and population displacement, resulting in significant excess mortality" (Toole 1995). Emergency responses following a disaster are primarily concerned with the surviving population rather than those killed in the disaster. Therefore, immediate action, seconds after a disaster has struck, should focus on life-saving activities. For example, in the event of an earthquake, rescuing people surviving under rubble and collapsed buildings should be prioritised rather than the evacuation of dead people. Post-disaster, the responses should address the need of the population directly affected by the disaster, i.e. injured and displaced people. The need for emergency responses to continue after an occurring disaster is assessed based on certain indicators, such as excess mortality, an indicator that is constantly monitored in emergencies.

One commonly used parameter linked to excess mortality is Crude Mortality Rate (CMR). CMR reflects the health status of the emergency-affected population (CDC 1992; Burkholder

& Toole 1995) and furthermore, relates to the number of deaths in a specified population over a specified period (Thomas & Thomas 2004). CMR has been widely used as measurement tools in complex emergencies to define phases of emergency like 'emergency phase' (CMR > 1 per 10,000 persons per day) and 'post-emergency phase' (CMR<1per 10,000 persons per day) (Spiegel *et al.* 2001; Thomas & Thomas 2004).

There is evidence that excess mortality following a disaster may not be directly caused by the disaster itself, but rather happens as a result of contracting diseases while staying in the displacement area. A recent study on the cause of deaths in Darfur, Sudan – a complex emergency case from prolonged conflict - highlighted that the majority of deaths occurred not due to violence but due to diseases that were contracted as the result of overcrowding and unsanitary conditions in displacement camps (Degomme 2011). For natural disasters, it was concluded that they are not associated with diseases outbreak when they do not result in massive displacement (Watson *et al.* 2007; Kouadio *et al.* 2011). Thus, excess mortality, as well as morbidity following disasters is closely associated with the health status of displaced people during displacements.

1.2 State of public health in displacements

Displaced people are situated in displacement centres, emergency shelters, public utilities, or are hosted by other surviving households. These locations are not prepared to cope with a sudden influx of a large group of people. Hence, it results in displaced people living in temporary settlements or camps with over-crowding and rudimentary shelters, inadequate safe water and sanitation, and increased exposure to disease vectors.

Specific observations indicated that the highest excess morbidity and mortality regularly occurs during the acute phase of an emergency, when relief efforts are in the early stage (Toole & Waldman 1990; Connolly *et al.* 2004). During this phase, deaths were up to 60 times the CMR when compared with non-refugee populations in the country of origin (Toole & Waldman 1990). In general, displacement increases these CMRs to at least double normal baseline rates in the population prior to any displacement activity (Thomas & Thomas 2004). Additionally, the high morbidity and mortality rate still occurs when the displacement continues. In protracted and post-conflict situations, populations may have high rates of illness and mortality due to the breakdown of health systems, flight of trained staff, failure of existing disease control programmes and destroyed infrastructure (Michelle Gayer 2007). These populations may be more vulnerable to infection and disease because of high levels of under-nutrition or malnutrition, low vaccine coverage, or long-term stress (Michelle Gayer 2007).

The major reported causes of death of refugees and internally displaced populations have been those same diseases that cause high death rates in non-displaced populations in developing countries, i.e. malnutrition, diarrheal diseases, measles, acute respiratory infections (ARIs), and malaria (Toole & Waldman 1988; Toole & Waldman 1990; CDC 1992). A longer list of displacement associated infectious diseases from more recent assessments includes diarrheal diseases, acute respiratory infections, malaria, leptospirosis, measles, dengue fever, viral hepatitis, typhoid fever, meningitis, in addition to tetanus and cutaneous mucormycosis (Kouadio *et al.* 2011). Amongst those infectious diseases, diarrheal diseases are the major contributors to overall morbidity and mortality rates following a disaster (Connolly *et al.* 2004; Waring & Brown 2005; Kouadio *et al.* 2011).

The World Health Organisation (WHO) defined 'diarrhea' or 'diarrhoea' as the passage of 3 or more loose or liquid stools per day, or more frequently than is normal for the individual. It is usually a symptom of gastrointestinal infection, which can be caused by a variety of bacterial, viral and parasitic organisms. Rotavirus and *Escherichia coli* (*E. coli*) are the two most common causes of diarrhoea in developing countries. Norwalk-like viruses, *Campylobacter jejuni*, and cytotoxigenic *Clostridium difficile* are seen with increasing frequency in developed areas; and moreover, Shigella, Salmonella, Cryptosporidium species and *Giardia lamblia* are found throughout the world (Guerrant *et al.* 1990). Following a disaster event, in a complex emergency situation, humanitarian agencies use WHO's classification of clinical diarrhoea to distinguish the many types of diarrheal diseases. In this regard, there are three types of clinical diarrhoea:

- Acute watery diarrhoea – lasts for several hours or days, and includes cholera
- Acute bloody diarrhoea - also called dysentery; and
- Persistent diarrhoea – lasts for 14 days or longer.

Diarrheal diseases are caused by intestinal based pathogens which are micro-organisms such as those transmitted via the faecal-oral route, which are closely associated with contaminated water supplies and food, particularly of faecal contamination water and food supplies, in addition to inadequate sanitation facilities. However, it is important to note that there are other diseases, although with little or no diarrhoea symptoms but transmitted similarly through faecal-oral contamination of water and or food, such as leptospirosis and hepatitis. Thus, these diseases are categorised together with diarrheal diseases such as 'waterborne diseases' (Waring & Brown 2005).

A living condition in unsanitary overcrowding locations, lack of clean water and safe sanitation, is a situation commonly experienced in displacements following a disaster and favours the spread of diseases more rapidly. The key measure to diarrhoea prevention is developments in access to clean water and safe sanitation, alongside behaviour changes towards hygiene practices and the clinical intervention of vaccination.

1.3 Emergency sanitation and urban sanitation

Regarding the humanitarian response context, the three sectors of water, sanitation and hygiene promotion are grouped into one cluster termed WASH, an acronym for water, sanitation and hygiene. The three sectors are grouped because of their close association with each other. Clinical intervention falls under the Health Cluster. Previously, before the realisation of a need to enhance behaviour change towards better hygiene practice, hygiene promotion was absence, leaving the cluster with only two sectors; specifically, water and sanitation.

When comparing water to sanitation, there are major discrepancies in relation to the two sectors, primarily attributed to the absence or minimal demand in the sanitation sector. While the demand for water has always been articulated, demand for sanitation was vaguely understood. Regarding the development context, availing sanitation for the sake of health improvement was scarcely the primary objective, compared to other benefits such as privacy, security, convenience, status, a reduction in flies and smell, and generally improved cleanliness (Scott *et al.* 2003). The same reality is transcended to the lack of provision of sanitation facilities in emergency settings where great health risks required attention resulting in low priority for sanitation programmes; hence, a lack of funding investments in sanitation, and consequently low interest in technology development for emergency sanitation. The focus on water supply in emergencies has made sanitation a forgotten area, repeatedly resulting in a sanitary disaster threatening the very same health objectives which a clean water supply aims to address (Johannessen *et al.* 2012). While there have been a lot of technological solutions developed to assist water provision in an emergency situation, technical options for sanitation remain limited. Agencies and donors are generally more willing to fund expensive water treatment units (which are regularly high-tech and can easily be shipped in one container) than to make the expenditure for sanitation systems – which are also less attractive in terms of media coverage (Andy Bastable, 2011 – as cited in Johannessen *et al.* (2012))

Emergency sanitation shares similarities of context with urban sanitation in developing countries. The type of sanitation system is commonly on-site sanitation with disintegrated elements of conveyance, treatment and final disposal. The context similarity of being in densely populated areas is also shared by emergency sanitation and urban sanitation. Figure 1-3 illustrates faecal sludge management in Dhaka, Bangladesh.

Similarly, even further limited sanitation services prevail in emergencies generating much higher public-health risks. Unsanitary living conditions in densely populated locations, with minimum availability of sanitation facilities, make public health risks more prominent than in common urban settlements.

1.4 The Bill & Melinda Gates Foundation Pro-Poor Sanitation Project framework

This research is one of 3 doctoral research studies under one research theme, i.e. 'Emergency sanitation following natural and anthropological disasters.' This theme is part of a larger research project (OPP1029019) that aims to stimulate local innovation on sanitation for the urban poor in Sub-Saharan Africa and South-East Asia. Presently the project is using the name 'Sani-UP' (Sanitation for urban poor). The Sani-UP project is funded by the Bill & Melinda Gates Foundation (BMGF).

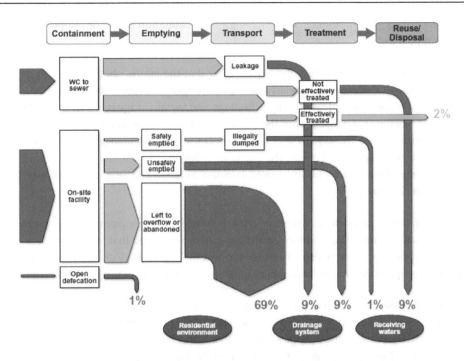

Figure 1-3 Faecal waste flows in Dhaka Bangladesh (Peal *et al.* 2014)

BMGF believes that innovation is the key to improving the world. Therefore, the foundation has been supporting innovations that have high impacts, making lasting changes to improve the lives of people suffering from hunger and extreme poverty. This research theme that deals with emergency sanitation, falls under the category of water, sanitation and hygiene (WASH) and emergency-response, which are two principal areas under BMGF's core global development programme.

1.5 Research gaps in emergency sanitation

In June 2012, a workshop on emergency sanitation was organised by WASTE – a Netherlands based non-governmental organisation concerned about creating opportunities in waste utilisation. The workshop was held at IHE Delft Institute for Water Education (formerly UNESCO-IHE), the Netherlands. It was attended by practitioners, relief workers, researchers, as well as representatives from the sanitation-related industries. It was an interaction opportunity for relief workers to share their problems and requirements to academia, researchers and supply industries. Vice versa, it was a forum for academia, researchers and supply industries to introduce their current innovations and products.

Following the workshop, an analysis was made of the research gaps. Those gaps were mapped and group into three topics i.e. (1) Reviews, analytical work and support tool developments;

(2) Technology developments; and (3) Testing and validation of the developed technologies and support tools.

It was decided when analysing the gaps further that the efforts should focus on technology innovations, as there are gaps related to technical options in the entire sanitation chain, i.e. collection/containment, de-sludging technology and faecal-sludge treatment prior to safe final disposal. Therefore, a large part of the research was directed towards technology innovation endeavouring to fill those gaps and towards a faecal sludge management system that takes care of the entire sanitation chain. The choice of technology to research ranges from basic (e.g. vermi-composting treatment and raised latrines) to advanced (e.g. microwave technology, membrane bioreactor, etc.).

However, more important issues beyond technology development were discovered. A preliminary emergency-sanitation technology review suggested that besides the lack of ready-to-use technology, there is also a need for support tools, such as decision-making packages to select appropriate sanitation technology to be used in different emergency scenarios.

Reviewing case studies is a way to identify gaps and validate analyses, and is identified as one of the research topics in the analysis of the gaps. Validation, application and/or field testing of innovated sanitation technologies have been lacking. It was learned that there are innovated technologies that have the potential to be used during emergencies but were not trialled in field testing or did not go through any external reviews. Relief agencies as main customers of emergency-sanitation products, confirmed that they would not use or purchase a product without a guarantee that the product is fail-proof. This remark is the basis of the decision to have all prototypes and products developed in this research tested or validated.

Besides the fail-proof guarantee, easy-deployment is another criterion for an emergency-sanitation product. This criterion includes aspects such as the size of the product, the use of light-weight material, foldable, modular, spare-parts that can be universally sourced and quick-construction. Some sanitation technologies have been often found to be successful in the experimental development stage but were found to be bulky and heavy, requiring special spare-parts and moreover, caused difficulties when being transported to emergency sites. It appeared necessary to design prototypes as deployable kit to satisfy the easy-deployment criterion.

This PhD research was directed to focus on a decision support system (DSS), improved raised latrines and a sanitation business model which includes the approach of the emergency sanitation DSS. The improved raised latrine subsequently became part of the eSOS system (emergency sanitation operation system), which later became known as eSOS Smart Toilet.

1.6 Scope of the study

Based on the analysis of the gaps, this particular research focuses on the development of a decision support tool and sanitation business model as software innovations, as well as hardware innovation relating to the development of an eSOS Smart Toilet from the design phase until field-testing of the experimental prototype.

1.7 Research objectives

The main objective of this research is to contribute to improving the quality of sanitation responses during a humanitarian crisis by way of technological and operational innovations, paying attention to the entire faecal sludge management service provision chain, facilitating the provision of safe sanitation in emergencies. The form of the technological innovations were specified as decision-making support tools to plan sanitation systems in emergencies and beyond. The research furthermore focused on to invent a smart toilet aiming to improve sanitation management in emergencies.

The specific research objectives are as follows:
1. To critically evaluate the planning for effective sanitation systems and contribute to its improvement by development of a software-based tool for general emergency sanitation technology selection;
2. To better understand toilet usage under emergency conditions (in this case – stabilization phase, instead of immediate emergency phase)
3. To assess the applicability of a smart toilet under real use in an emergency settlement;
4. To evaluate the performance of functionalities embedded in a smart toilet;
5. To develop and critically evaluate a developed sanitation business model

1.8 Outline of the thesis

This thesis consists of ten chapters. This first chapter provides the general introduction and justification of the focus of the research. Two chapters are dedicated to software development, each one for decision support systems for emergency sanitation and sanitation business model software. The middle section chapters discuss the findings obtained from developing and field testing an eSOS smart toilet. The final chapter concludes and summarises the findings from this research and provides an outlook on the topic.

- **Chapter 1** provides introduction, background context and the rationale of the research topics
- **Chapter 2** describes the rationale of the eSOS concept and elements of the eSOS sanitation chain.
- **Chapter 3** describes the development of the eSOS smart toilet, adopted ICTs (information communication technology) in its features and its deployment to the field-testing site in the Philippines.
- **Chapter 4** reports the research findings from the field testing of the eSOS smart toilet, focusing on the obtained usage data and its application to design refinement of the smart toilet.

- **Chapter 5** reports the effectiveness of the water treatment unit in the eSOS toilet and waste streams quality analysis to recommend suitable treatment options or a disposal management plan.
- **Chapter 6** evaluates the effectiveness of the UV-C light featured in the eSOS toilet to assist surface disinfection for self-cleaning of the toilet, reducing the burden of manual cleaning, as well as guaranteeing the cleanliness of the toilet at every visit.
- **Chapter 7** assess the user's acceptance of the eSOS toilet, as well as gaining opinions from the residents of the testing site regarding the toilet design refinement.
- **Chapter 8** describes the developed decision support system for the provision of emergency sanitation.
- **Chapter 9** describes the developed business model software that serves as a decision support system for general sanitation that includes the DSS model described in Chapter 8.
- **Chapter 10** provides general discussion of the findings discussed in each chapter and an outlook recommending further researches and improvements.

References

Burkholder B. T. and Toole M. J. (1995). Evolution of complex disasters. *The Lancet* **346**(8981), 1012-5.

CDC (1992). Famine-affected, refugee, and displaced populations: recommendations for public health issues. *MMWR. Recommendations and reports : Morbidity and mortality weekly report. Recommendations and reports / Centers for Disease Control* **41**(RR-13).

Connolly M. A., Gayer M., Ryan M. J., Salama P., Spiegel P. and Heymann D. L. (2004). Communicable diseases in complex emergencies: impact and challenges. *The Lancet* **364**(9449), 1974-83.

Degomme O. (2011). Mortality in Darfur: Lessons for Humanitarian Policy. *MICROCON Policy Briefing*(7).

Guerrant R. L., Hughes J. M., Lima N. L. and Crane J. (1990). Diarrhea in Developed and Developing Countries: Magnitude, Special Settings, and Etiologies. *Review of Infectious Diseases* **12**(Supplement 1), S41-S50.

IFRC (2010). World disasters report 2010: focus on urban risk. http://www.ifrc.org/en/publications-and-reports/world-disasters-report/wdr2010/.

Johannessen A., Patinet J., Carter W. and Lamb J. (2012). Sustainable sanitation for emergencies and reconstruction situations. In: *Factsheet of Working Group 8*, Sustainable Sanitation Alliance (SuSanA).

Kouadio I. K., Aljunid S., Kamigaki T., Hammad K. and Oshitani H. (2011). Infectious diseases following natural disasters: prevention and control measures. *Expert Review of Anti-infective Therapy* **10**(1), 95-104.

Michelle Gayer D. L., Pierre Formenty, Maire A. Connolly (2007). Conflict and Emerging Infectious Diseases. *Emerging Infectious Diseases* **13**(11).

NatCatSERVICE (2015). Münchener Rückversicherungs-Gesellschaft, Geo Risk Research. In: *NatCatSERVICE, Natural catastrophes*.

Peal A., Evans B., Blackett I., Hawkins P. and Heymans C. (2014). Fecal sludge management: a comparative analysis of 12 cities. *Journal of Water, Sanitation and Hygiene for Development* **4**(4), 563-75.

Scott R., Cotton A. and Govindan B. (2003). Sanitation and the Poor. In, WELL.

Spiegel P. B., Sheik M., Woodruff B. A. and Burnham G. (2001). The Accuracy of Mortality Reporting in Displaced Persons Camps During the Post-emergency Phase. *Disasters* **25**(2), 172-80.

Thomas S. L. and Thomas S. D. (2004). Displacement and health. *British Medical Bulletin* **69**(1), 115-27.

Toole M. J. (1995). Mass population displacement. A global public health challenge. *Infectious Disease Clinics of North America* **9**(2), 353.

Toole M. J. and Waldman R. J. (1988). An analysis of mortality trends among refugee populations in Somalia, Sudan, and Thailand. *Bulletin of the World Health Organization* **66**(2), 237.

Toole M. J. and Waldman R. J. (1990). Prevention of excess mortality in refugee and displaced populations in developing countries. *JAMA: The Journal of the American Medical Association* **263**(24), 3296-302.

UNHCR (2015). *World at war : UNHCR global trends : forced displacement in 2014*. Office of the United Nations High Commissioner for Refugees, Geneva.

UNHCR (2016). Global trends: Forced displacement in 2015. In, United Nations High Comissioner for Refugees, Geneva.

Waring S. C. and Brown B. J. (2005). The Threat of Communicable Diseases Following Natural Disasters: A Public Health Response. *Disaster Manag Response* **3**(2), 41-7.

Watson J. T., Gayer M. and Connolly M. A. (2007). Epidemics after natural disasters. *Emerging infectious diseases* **13**(1), 1.

Newly arrived displaced people in South Darfur, Sudan in 2010 (Photo by F. Zakaria)

eSOS™ - emergency Sanitation Operation System

This chapter is based on:

Brdjanovic D., Zakaria F., Mawioo P. M., Garcia H. A., Hooijmans C. M., Ćurko J., Thye Y. P. and Setiadi T. (2015) eSOS™ – emergency Sanitation Operation System, *Journal of Water, Sanitation and Hygiene for Development* 5 (1), 156–164 (*IF 0.8*)

Abstract

This chapter presents the innovative emergency Sanitation Operation System (eSOS) concept created to improve the entire emergency sanitation chain and provide decent sanitation to people in need. The eSOS kit was described including its components: eSOS smart toilets, an intelligent excreta collection vehicle-tracking system, a decentralized excreta treatment facility, an emergency sanitation coordination center, and an integrated eSOS communication and management system. Further, the chapter deals with costs and the eSOS business model, its challenges, applicability, and relevance. The first application, currently taking place in the Philippines will bring valuable insights on the future of the eSOS smart toilet. It was expected that eSOS would bring changes to traditional disaster relief management.

Key words: emergency, faeces, sanitation, technology, toilet, urine

2.1 Emergency sanitation

In general, an emergency can be considered to be the result of a man-made and/or natural disaster, whereby there is a serious, often sudden, threat to the health of the affected community which has great difficulty in coping without external assistance. Emergency sanitation intervention is a means of promoting best management practice to create a safer environment and minimize the spread of disease in disaster-affected areas, and of controlling and managing excreta, wastewater, solid waste, medical waste, and dead bodies. In June 2012, an international emergency sanitation conference was hosted by IHE Delft where more than 200 experts from relief agencies, governments, academia and industry gathered and discussed emergency excreta management and public health. It was confirmed that (i) emergency-specific sanitation is not at the forefront of the scientific community, (ii) current solutions are in most cases technologically and economically suboptimal, (iii) there is, in general, insufficient communication between key stakeholders, (iv) academia and practitioners are insufficiently involved, (v) emergency sanitation (technological) development is often associated with drivers such as humanitarian aid agencies or the army, (vi) emergency water supply is given much more attention than sanitation, and (vii) the smart innovative emergency sanitation management (and governance) system is lacking.

This concept aims to address these deficiencies and provide sustainable, innovative, holistic, and affordable sanitation solutions for emergencies (such as floods, tsunamis, volcano eruptions, earthquakes, wars, etc.) before, during, and after a disaster.

2.2 eSOS™

The abbreviation eSOS stands for the innovative 'emergency Sanitation Operation System' concept (Brdjanovic *et al.* 2013). This concept addresses the tire emergency sanitation chain (Figure 2-1). It is based on a balanced blend of innovative sanitation solutions and existing information technologies adapted to the specific conditions of emergency situations and in informal settlements. The central points of the system are the reinvented smart emergency toilet and the innovative decentralized treatment of excreta, embedded in an intelligent emergency sanitation operation system. Information and communication technologies have a unique opportunity to assist following disasters because the core of any emergency management effort is integration, sharing, communication, and collaboration, things that stakeholders involved embrace and promote.

2.3 eSOS™ kit

The eSOS is based on different system components integrated into an easily deployable emergency sanitation kit consisting of hardware and software components. The software components include the communication chain by controlling the mobile network and the Local Area Network (LAN)/Wide Area Network (WAN) simultaneously. The routing application supports receiving data messages –from General Packet Radio Service (GPRS) networks and the SMS channel – from large quantities of Global System for Mobile Communications (GSM) and Code Division Multiple Access (a radio channel access method) units at the same time. Alternatively, a non-GSM-based system can be applied to disaster sites which are not covered by a GSM network (e.g. remote refugee camps) or are temporarily without GSM coverage due to a disastrous natural event. Also, a portable navigation system is used to supplement excreta collection vehicle-tracking. Geographic Information System (GIS) maps and data, as well as other interactive and public domain information, are used and combined into this integral eSOS, such as digital orthophotography, digital terrain maps, land-use maps, sanitary points of interest, and population density maps. It is all combined in user-friendly software with an intuitive graphic interface to allow rapid advance to the expert user level. The components of the eSOS are smart toilets, intelligent excreta collection vehicle-tracking systems, decentralized excreta treatment facilities, emergency sanitation coordination centers, and integrated eSOS communication and management systems.

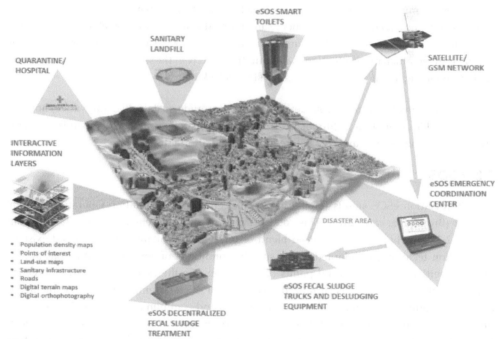

Figure 2-1 eSOS concept components (Brdjanovic *et al.* 2013)

2.4 eSOS™ Smart Toilets

Sanitation facilities usually provided by relief agencies and armies have additional specifications and requirements in comparison to those regularly used in other settings. The eSOS Smart Toilets have the following characteristics: they are stackable and lightweight, fit a Euro-size pallet, are made of durable materials, are easy to wash and clean, are easy to empty, require minimum maintenance, are raised above the ground, do not require any excavation to install, allow more frequent use, provide excellent value for money, are easy and safe to use, provide privacy, are easily deployable, give a sense of dignity to users, look great and invite usage, etc.

Beside these aspects, the eSOS concept addresses the 'smartness' of the emergency sanitation toilet by incorporating unique (either as 'built-in' or 'add-on') features such as: interchangeable squatting pans or sitting toilet, delivered as a urine diversion dry toilet or flush toilet, safe and easy-to- empty storage of urine and faeces, fully solar-powered with up to 7 days energy independency, GSM-based communication, GPS-based tracking, real-time information on occupancy, volume of urine collected, volume of service water and gray water and UV interior disinfection, nano-coated interior, smart card reader entry system, SOS panic button, smart software for monitoring, data collection and optimization, etc.

Beside smart data collection and communication, the eSOS toilet is subject to technological innovations from the sanitary engineering perspective. It is a urine diversion toilet with separate collection (and treatment) of urine and faeces, with both 'dry' and 'wet' sanitation options. It is important to note that the eSOS toilet is not designed as an on-site treatment unit due to its high-frequency use and limited storage capacity. The rule of thumb applied by relief agencies of a maximum of 50 users per day will be evaluated during field testing and verified later by data gathered from eSOS toilets to be installed worldwide operating under different conditions. At the moment the capacity of the urine tank and faeces tank in one unit is 80 L each. This arrangement should be revised following feedback from experimental testing. It allows for an emptying interval of individual units of about once a week for a 'dry' toilet. In case the 'community' type of arrangement is applied (several toilets in a cluster), a common larger storage tank will replace individual units allowing for significantly larger storage, more frequent use and less frequent emptying. Longer retention times and ongoing processes in stored faeces and urine will be taken into account in the design of such clustered applications at a later stage of the development of the eSOS system. Of course, the situation will change in the case where continuous or intermittent water supply system and sewer system are available where the 'wet' option may well be applied. As the urine tank makes up part of the toilet body, it will be possible to empty it only on-site by gravity or by a vacuum truck. For faeces evacuation, several emptying options will be possible: by vacuum truck, by replacing a full tank with an empty one, and by several ways of emptying the tank manually on-site (e.g. there is an analogy with vacuum cleaner bags).

Owing to specific emergency requirements, its innovative light-weight, stackable toilet structure is proposed to be made of recycled biodegradable materials (like bio-plastic made from potato skins). Options for both on-site and centralized treatment (and their combination) of urine and faecal sludge is also investigated. Packed, a complete toilet kit occupies a volume of 2 m^3 which will allow for compact and cheaper shipping (a toilet fits one standard pallet). Owing to its modular set-up, it will be possible to quickly and simply install the toilet on the spot. Simplified instructions on how to install and use the toilet will be provided with the kit. Each part of the toilet is unique and can only be assembled in one way to avoid confusion. In the near future, possibilities to produce toilets locally shall be explored, also using local materials. However, in general, it will not be possible to produce these toilets at the disaster location. The present version of the toilet allows for its usage by both children and adults and women and men.

In addition, several variations of the eSOS toilet were produced in a later stage of development to account for different settings and conditions and user groups including elderly citizens, people with disabilities and the injured. Development of the eSOS smart toilet was carried out in two steps, namely: design, manufacturing, and field testing of the 'experimental toilet' (Figure 2-2) and based on the feedback from field testing and relief practitioners, the 'design vision toilet prototype' (Figure 2-3) will be manufactured. Shortly after its development, the experimental eSOS toilet was field-tested at a disaster site in Tacloban City in the Philippines. During several months of testing, an extensive research program was executed; which revealed novel information on the use of a toilet in an emergency setting (see the field testing results from Chapter 4 to 7).

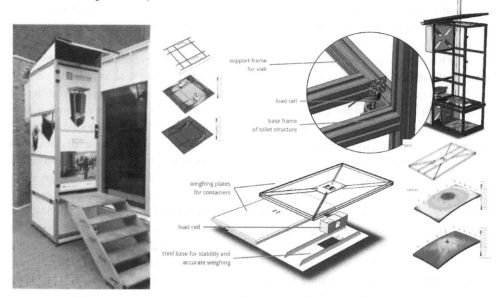

Figure 2-2 eSOS smart experimental toilet. The toilet's structure and electronic features have been subject to extensive testing during the manufacturing phase (Photo: D. Brdjanovic; drawings: Flex/design)

2.5 Intelligent tracking system for excreta collection vehicles

In emergency situations, due to high traffic and load to toilets, frequent emptying (of relatively fresh urine and faeces) is required, which consequently creates demand for well-organized logistics for excreta collection, a feature which is regularly lacking during, by definition, rather chaotic emergency circumstances. As an emergency may last for days, months, and sometimes years, the issue of excreta management and logistics becomes extremely important in sustaining the emergency sanitation chain. For example, in the first few months after the 2010 earthquake disaster in Haiti, the costs for de-sludging toilets and latrines exceeded USD 0.5 M. The eSOS envisages the use of GPS- (or satellite-) based communication infrastructure; e.g. a real-time GPS vehicle-tracking system, where each truck and/or each trailer/cistern is equipped with 'easy-to-install' GSM/GPS sensor/card (similar to those supplied with or to eSOS toilets), which allows 24/7 information of the position (and route) of each toilet-emptying vehicle. This information 'feeds' the advanced, commercially available, vehicle tracking system, and software and on-board location-based analysis, which processes data and provides much useful information (e.g. route optimizer, total amount of urine and faeces collected per day, disposal location, etc.) to the user in the emergency sanitation coordination centre.

Figure 2-3 eSOS Smart Toilet design vision prototype (images: Flex/design)

Efforts will be made to rapidly update the navigation maps with the most recent information regarding the disaster event (accessibility of roads, bridges, tunnels, etc.) and isolate sections with limited or no traffic, most likely based on physical site inspection with the eventual support of updated satellite images that can be purchased on demand as an add-on feature of the integrated eSOS.

2.6 Excreta treatment facility

Three distinctive emergency sanitation phases are generally adopted in the work of relief agencies, namely: (i) phase 1 of duration up to 2 weeks, where the main mean for sanitation provision is individual, mass-production, inexpensive kits (like biodegradable PeePoo bags), (ii) phase 2, lasting upto a few months, where substantial sanitation hardware components are supplied to the disaster site (like individual portable toilets or clusters of those, and de-sludging equipment and vehicles), and (iii) phase 3, which can last from several months up to a few years or longer where more (semi)permanent sanitation hardware is supplied such as community-based toilets and (mobile) excreta treatment facilities (more sophisticated package/containerized plants or, sometimes, on-site/land-based simplified solutions). Comparatively much higher load (increased usage per toilet), consequent requirements for more frequent emptying, and different faecal sludge characteristics (fresh biologically non-stabilized sludge and fresh non-hydrolyzed urine, with higher public health risk), are distinctive, but often overlooked features of emergency sanitation. Therefore, the current management practices in emergency sanitation need a thorough revision and re-thinking, especially from the treatment perspective, as to this aspect 'the business as usual' approach is applied, often not being fully aware of specific technological and social key issues of concern. Although many standard options for faecal sludge management in general already exist, their application in emergency situations is not well understood and is often lacking. To address these deficiencies, Sanitation Team of IHE Delft conceptualized, designed, manufactured, tested and applied on a pilot scale an innovative, compact, and efficient treatment of emergency sanitation faecal sludge, including (separate) treatment of urine, by physical–chemical treatment-based technologies (e.g. microwave technology and/or dewatering/drying) with specific attention on public health (epidemiologic) aspects and safe disposal of treatment residuals (Figure 2-4).

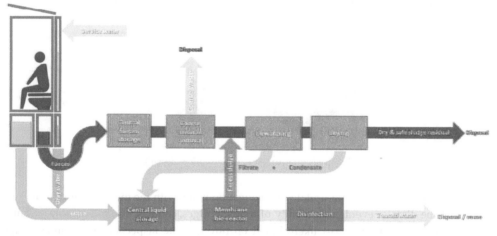

Figure 2-4 eSOS™ excreta treatment concept (Brdjanovic *et al.* 2013)

The novel faecal sludge microwave – brand system is popularly called The Shit Killer, is considered a promising solution for fast pathogen inactivation and sludge dehydration (Mawioo *et al.* 2016a; Mawioo *et al.* 2016b; Mawioo *et al.* 2017). For the emergency sanitation, also due to economies of scale, it is more appropriate to apply treatment solutions in decentralized on-site settings, rather than solving the excreta issue at the level of individual toilets. However, the decentralized technology being developed here is equally applicable (with some modifications) at small scales as well. After initial testing in Slovenia, the installation will be subject to field testing in Jordan using excrement collected from a refugee camp.

2.7 Emergency sanitation coordination centre

The emergency sanitation coordination centre is the heart of eSOS concept, and should be located either on-site or at any remote location outside the disaster area. It has a high degree of automation and requires an operator/coordinator.

For the immediate response, and if no skilled operator is available, remote operation is possible by an expert operator located outside the disaster area. The coordination center will be equipped with the central information processing unit (laptop or tablet) which will contain all necessary software and will receive and process all relevant information for the eSOS in the cloud. If on-site, the centre will be responsible for physically inspecting and verifying some of the key information collected by remote sensing and making sure that the correct information is used (e.g. accessibility of roads and correct location of existing sanitary infrastructure used in emergency, like sanitary landfill, (decentralized wastewater treatment plant, empty industrial storage tanks, and in extreme cases, temporary discharge points to open environment, etc.).

2.8 eSOS™ operation

Based on information, such as population density maps or real-time population tracking using mobile telephony and other information automatically acquired from the disaster area, in combination with the user-entered information, the operator will have a rather good understanding of where to position the emergency aids. The number of sanitation units deployed will be initially determined using rules of thumb (e.g. up to 50 people per toilet per day), but the application of eSOS will very soon provide practical feedback on these rules as much more (new) data will become available. In addition, based on the existing population density and real-time information on the population migration using mobile telephone signals, the optimal locations (density) of available sanitation units (Pee-Poo bags, for example, for the immediate response, followed by the supply of emergency toilets) will also be determined.

In the case of pre-fabricated eSOS toilets, they will automatically report their location to the central system (coordination centre) and will appear on the interactive disaster area map as such. In cases, where the toilet is not equipped with an eSOS kit, it can easily be retrofitted by rapid installation of the necessary sensors and electronic equipment. Also, already existing units can be upgraded with this equipment, so that the entire emergency sanitation facilities are tagged and included in the network.

The second step is to equip the excreta collection vehicles with the tracking electronic and navigation equipment. This can be done very easily and quickly by installing the removable equipment preferably inside the driver's cabin.

The third step is to mobilize the central data collection and processing unit with all the required software necessary for the operation of the eSOS and to ensure that the Internet connection or access to a cloud computing/server facility via a satellite connection is available.

After the system is up and running, the operator can use all above-described features to apply the eSOS in a rapid, more efficient, and economic fashion, with increased confidence. The eSOS system is designed as a stand-alone application, refined at the operator's emergency center. It enables the definition of the required procedure for each stage in an emergency and to react to every call within the shortest period of time. It also enables the local operator to define the unit's parameters according to both the customer's and local network demands and to create (daily, weekly, monthly) reports with statistics and performance indicators.

The authors and funding agencies disseminated all useful feedback from the practical applications in separate papers (Zakaria *et al.* 2016; Zakaria *et al.* 2017; Zakaria *et al.* 2018) and through other methods of communication. These will soon after be translated into a user manual or operational guide as a part of the eSOS emergency kit.

2.9 Costs and eSOS financial flow model

The current conceptual state of the development of the system does not yet allow being accurate where the costs are concerned for the following reasons. The costs and benefits will depend on many factors where the production and operations costs combined with the location-specific conditions and scale of disaster and number of people affected/served will determine the total financial picture.

As both emergencies and disasters have a high degree of uncertainty associated with them and since disasters can strike anywhere in the world at any given moment and given that emergencies have different characteristics and phases, it makes the current application of standard sanitation financial models inadequate and only remotely accurate and useful. As a part of the eSOS concept, the development of a holistic business model is demanded and has therefore been developed with extended boundaries to capture aspects traditionally difficult to estimate (thus often neglected) but essential to such an assessment, such as costs (and benefits) related to public health (hospitalization, absence from work, productivity, temporary or permanent disability and casualties, quality of life, dignity, safety, etc.). The model is interactive, adaptable to local conditions and specifics of emergency sanitation, and also includes costs for production (e.g. rotational molds, materials, 'add-ons', labor, etc.), costs for storage, transport and erection, costs for operation and maintenance, and costs for eventual deployment, depreciation, etc. It is expected that in the majority of emergency situations, the additional unconventional features and elements of the system and associated costs will be at least compensated for if not overwhelmed by the benefits that such a system can bring. The new eSOS financial flow model included feedback from major relief agencies and all other key players in emergency relief, also included demonstrations with detailed costs analyses, and is verified on several case studies that shall provide more confidence in using it (See Chapter 9). The financial flow model is in the public domain (web page).

2.10 Challenges

The eSOS confirms the rule that one involved in the process of moving from invention to innovation faces a number of challenges such as how to make a product which will match its purpose at an affordable price with maximized benefits. The eSOS components are designed to satisfy specific requirements of relief operations regarding materials, durability, resistance to theft and misuse, demands of users, environmental and public health, cultural and social features of societies, and must also be attractive to people so that they make use of it in the first place. Expectedly, the eSOS concept cannot possibly be a solution for each and every emergency situation and its future will depend on acceptance, affordability, effectiveness and efficiency of operations, and the extent to which the limitations will be overcome by further development and incorporation of the feedback from practical applications.

2.11 Applicability and relevance

The strength of eSOS is that it is addressing, improving, and making each component of the emergency sanitation chain smarter, taking care that innovations also take place at the level of the system. The eSOS system is globally applicable to a wide spectrum of emergency situations where external aid is needed for sanitation. The eSOS concept, with minor adaptations, can be made equally suitable for, but is not limited to (i) sanitation management under challenging conditions usually prevailing in urban-poor areas, such as slums and informal settlements, (ii) sanitation provision for visitors of major open-air events such as concerts, fairs, etc., and (iii) solid waste management.

So far, initial constructive and in general encouraging feedback from several parties, including the United Nations Children's Fund (UNICEF), United Nations Refugee Agency (UNHCR), Red Cross, Oxfam, Save the Children, Doctors without Borders (MSF) and OPEC Fund for International Development (OFID), has already been received. It is planned to have key players in relief provision more actively involved in the further development of the eSOS system.

Part of the research in the Philippines and other locations provided us with lessons and answers on how to ensure the uptake of the system. At the moment, the framework for how to commercialize the eSOS and build a business case for the new eSOS enterprise is drafted. It will also include important aspects such as after-sales services that will be very much dependent on the type of emergency, local conditions, culture, emergency setting, etc. The fate of eSOS in a post-disaster period will also be considered.

If the life returns to 'normal' and original infrastructure is recovered, the eSOS can be cleaned, dismantled and reused elsewhere as the system allows for it. In the case where new (semi)organized settlements are created, like refugee camps, the eSOS may remain there, given that a proper governance system and the business case are in place to make it sustainable, making the eSOS of more permanent character. In the case where the eSOS is used for non-emergency situations (events, etc.); it will be reused.

In the case of its use in informal settlements (slums), it will be of permanent character. Present design takes care as much as possible that the system is theft proof (the comment on theft and costs of eSOS came up often in social media).

The potential clients/end-users are relief agencies, municipalities, water and sewerage companies, solid waste companies, army, police, fire brigades, as well as private sector companies and water supply, and sanitation vendors.

The primary goal of eSOS is to save lives by providing an efficient and effective sanitation service during and after emergencies through minimizing the risk to public health of the most vulnerable members of society. The secondary goal is to reduce the investment, operation, and maintenance costs of emergency sanitation facilities and service as a pre-requisite for sustainable solutions, especially in the post-emergency period.

2.12 Concluding remarks

The innovative eSOS concept provides a sustainable, innovative, holistic, and seemingly affordable sanitation solution for emergencies before, during, and after disasters. eSOS does not only reinvent the (emergency) toilet and treatment facilities, but uses existing information and communication technology to bring innovation and potential cost savings to the entire sanitation operation and management chain, and most importantly, is expected to improve the quality of life of people in need.

2.13 Acknowledgements

The eSOS™ concept is developed under the project 'Stimulating local innovation on sanitation for the urban poor in Sub-Saharan Africa and South-East Asia' financed by the Bill & Melinda Gates Foundation. The eSOS™ smart toilet design concept is a joint effort of IHE Delft, FLEX/design and SYSTECH.ba. First eSOS smart toilet testing in the Philippines is supported by the Asian Development Bank and Bill & Melinda Gates Foundation. The eSOS™ concept is an invention of IHE Delft.

References

Brdjanovic D., Zakaria F., Mawioo P. M., Thye Y. P., Garcia H., Hooijmans C. M. and Setiadi T. (2013). eSOS® innovative emergency sanitation concept. In: *3rd IWA Development Congress and Exhibition*, Nairobi, Kenya.

Mawioo P. M., Garcia H. A., Hooijmans C. M., Velkushanova K., Simonič M., Mijatović I. and Brdjanovic D. (2017). A pilot-scale microwave technology for sludge sanitization and drying. *Science of The Total Environment* **601**, 1437-48.

Mawioo P. M., Hooijmans C. M., Garcia H. A. and Brdjanovic D. (2016a). Microwave treatment of faecal sludge from intensively used toilets in the slums of Nairobi, Kenya. *Journal of environmental management* **184**, 575-84.

Mawioo P. M., Rweyemamu A., Garcia H. A., Hooijmans C. M. and Brdjanovic D. (2016b). Evaluation of a microwave based reactor for the treatment of blackwater sludge. *Science of The Total Environment* **548**, 72-81.

Zakaria F., Ćurko J., Muratbegovic A., Garcia H. A., Hooijmans C. M. and Brdjanovic D. (2018). Evaluation of a smart toilet in an emergency camp. *International Journal of Disaster Risk Reduction* **27**, 512-23.

Zakaria F., Harelimana B., Ćurko J., van de Vossenberg J., Garcia H. A., Hooijmans C. M. and Brdjanovic D. (2016). Effectiveness of UV-C light irradiation on disinfection of an eSOS® smart toilet evaluated in a temporary settlement in the Philippines. *International Journal of Environmental Health Research* **26**(5-6), 536-53.

Zakaria F., Thye Y. P., Hooijmans C. M., Garcia H. A., Spiegel A. D. and Brdjanovic D. (2017). User acceptance of the eSOS® Smart Toilet in a temporary settlement in the Philippines. *Water Practice and Technology* **12**(4), 832.

3

eSOS™ Smart Toilet development history

3.1 Introduction

Inspired by participation in the emergency sanitation workshop in Delft in June 2012 and the Reinvent Toilet Fair at the BMGF premises in Seattle in August 2012, Prof. Brdjanovic came up with the idea of eSOS (emergency Sanitation Operation System) innovative concept. This concept addresses the entire emergency sanitation chain. It is based on the balanced blend of the innovative solutions and existing ICT technologies adapted to specific conditions of the emergency situations and conditions in informal settlement (Brdjanovic *et al.* 2015). The primary goal of the eSOS is to save lives by providing efficient and effective sanitation service during and after emergencies through minimizing risks to public health of especially vulnerable members of society. The secondary goal is to reduce investment, operation and maintenance costs of emergency sanitation facilities and services as prerequisite for the sustainability solutions, especially in the post emergency period. The eSOS system and smart toilets are globally applicable to wide spectrum of emergency situations where external aid is needed for sanitation. The eSOS concept, with minor adaptation, can be made equally suitable for and not limited to (i) sanitation management under challenging conditions usually prevailing in urban-poor areas, such as slums and informal settlements, (ii) sanitation provision to visitors of major open-air events such as religious gathering, concerts, fairs, sport events., and (iii) solid waste management. The central point in the eSOS concept is eSOS Smart Toilet. So, what is so special about this toilet?

The eSOS Smart Toilet has many unique features distinguishing itself from the products presently available on the market, just to mention some: interchangeable squat or pedestal type of toilet, delivered as urine diversion dry toilet or flush toilet, safe and easy to empty storage of urine and faeces, fully solar-powered with up to 7 days energy independency, GSM-based communication feature, GPS-based location tracking feature, real time information (occupancy, volume or urine collected, amount of faeces collected, volume of service water and grey water), UV interior disinfection, nano-coated interior, smart card reader entry system, SOS panic button, mini unit for water treatment, software for monitoring and optimization, etc.

The eSOS Smart Toilets are also stackable and lightweight, made to fit a Euro-size pallet, made of durable materials, easy to wash and clean, easy to empty, easy to wash and clean, easy to empty, require minimum maintenance, include interchangeable squatting pan or sitting toilet, raised above the ground, do not require any excavation to install, allow for higher frequency of usage, excellent value for money, easy and safe to use, providing privacy, easily deployable, giving sense of dignity to users, look great and invite usage etc.

The first serious steps towards the design of the eSOS toilet started late 2013 at the workshop in Delft where representatives of the eSOS team put together the basic design concept and agreed on functionalities and components that the eSOS toilet should have (Figure 3-1). At that meeting it was decided to split the toilet development into two overlapping phases, namely the development of the so called experimental eSOS Smart Toilet and the development of the eSOS Smart Toilet final vision prototype. The purpose of having the experimental toilet is to use it for development and proofing the concept with focus on the applied research, while the final prototype is the necessary step bridging the development and commercial production. The eSOS Smart Toilet in its final form is not expected to have all features and functionalities of the experimental toilet which are in the final prototype limited to the minimum required operational functionalities.

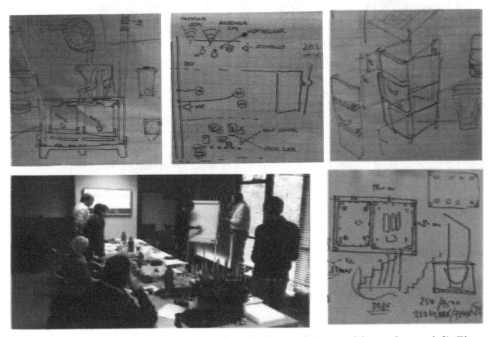

Figure 3-1 Workshop to conceptualize eSOS toilet: sketches, workshop participants (bottom left) (Photos: D. Brdjanovic)

3.2 The experimental eSOS Smart Toilet

Soon after the eSOS team meeting, the FLEX/design came up with the first concept design of the experimental toilet (Figure 3-2). This experimental unit is equipped with all components as visioned in the eSOS concept. These components are detailed in Figure 3-3. The development of the individual components of eSOS experimental toilet is described further in the text. Further development, detailed design and manufacturing of the experimental eSOS toilet took approximately 6 months. Finally, on 11 July 2014 the festive opening took place at premises of IHE Delft (Figure 3-4).

3.2.1 User interface

The experimental eSOS Smart Toilet has unique and innovative feature of having interchangeable squat and pedestal type of interface enabling it to be adapted to different social, cultural and religious environments (Figure 3-5). In both cases the toilet preserves the urine diverting feature (UDT). For this purpose two interchangeable toilet floors were constructed, one fitted with device developed by Sanergy (Nairobi, Kenya) and one with a commercially available pedestal.

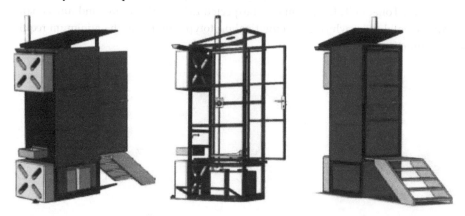

Figure 3-2 The conceptual design of the experimental eSOS Smart Toilet (Images: Flex/design)

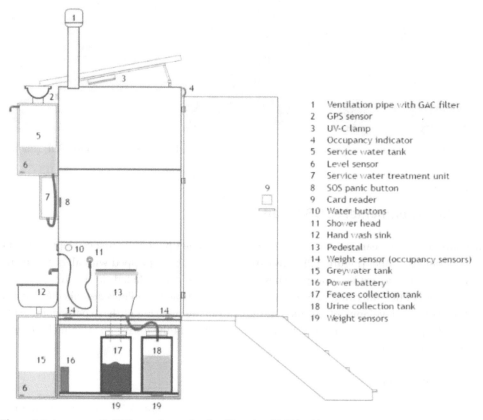

1   Ventilation pipe with GAC filter
2   GPS sensor
3   UV-C lamp
4   Occupancy indicator
5   Service water tank
6   Level sensor
7   Service water treatment unit
8   SOS panic button
9   Card reader
10 Water buttons
11 Shower head
12 Hand wash sink
13 Pedestal
14 Weight sensor (occupancy sensors)
15 Greywater tank
16 Power battery
17 Feaces collection tank
18 Urine collection tank
19 Weight sensors

Figure 3-3 Anatomy of eSOS experimental toilet (Drawing: F. Zakaria)

3.2.2 Safe and easy-to-empty storage of urine and faeces

As the eSOS Smart Toilet is an UDT (urine diverting toilet) with separate collection (and treatment) of urine and faeces (Figure 3-6), with both 'dry' and 'wet' sanitation options, it is important to note that the eSOS Smart Toilet is not designed as an onsite treatment unit due to its high-frequency use and limited storage capacity. The rule of thumb applied by relief agencies of a maximum of 50 users per day will be evaluated during field testing (Chapter 4) and verified later by data gathered from eSOS toilets to be installed worldwide operating under different conditions. At this stage, the capacities of the urine tank and faeces tank in one unit is 80-L each. This arrangement was revised after experimental testing (Chapter 4). It was estimated to allow for an emptying interval of individual units of about once a week for a 'dry' toilet. In case the 'community' type of arrangement is applied (several toilets in a cluster), a common larger storage tank will replace individual units allowing for significantly larger storage, more frequent use and less frequent emptying. Longer retention times and ongoing processes in stored faeces and urine will be taken into account in the design of such clustered applications at a later stage of the development of the eSOS system. Of course, the situation will change in case continuous or intermittent water supply system and sewer system is available where the 'wet' option may well be applied. As the urine tank makes part of toilet body it will be possible to empty it only onsite by gravity or by a vacuum truck. For faeces evacuation several emptying options will be possible: by vacuum truck, by replacing full tank with empty one, and by several ways of emptying the tank manually onsite (e.g. there is an analogy with vacuum cleaner bags).

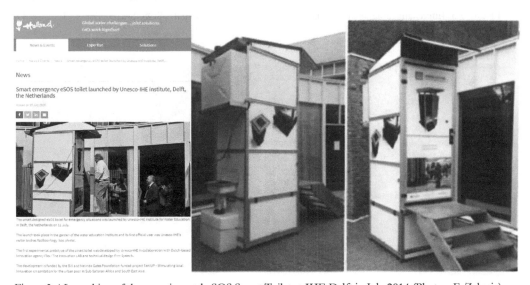

Figure 3-4 Launching of the experimental eSOS Smart Toilet at IHE Delft in July 2014 (Photos: F. Zakaria)

3.2.3 Energy supply

The energy to power eSOS Smart Toilet is provided by a solar panel mounted at the roof of the toilet and batteries located at the bottom part of the toilet (Figure 3-7). The amount of collected energy is sufficient to supply the toilet with energy for usual operation for at least 7 days. The level of power in the battery is possible to see in the monitoring software window.

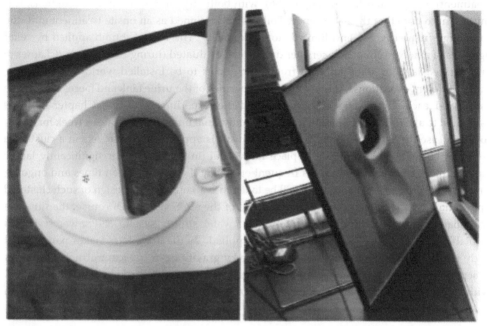

Figure 3-5 Interchangeable user interface used in the experimental eSOS Smart Toilet (Photos: D. Brdjanovic)

Figure 3-6 Evacuation of urine and faces tanks from experimental eSOS Smart Toilet during testing in Tacloban City, Philippines (Photos: F. Zakaria)

3.2.4 Communication and tracking

The eSOS Smart Toilet is integrated into an easy-deployable emergency sanitation kit consisting of hardware and software components. The software components include the communication chain by controlling the mobile network and the LAN/WAN simultaneously. The routing application supports receiving data messages - from GPRS networks and the SMS channel -from large quantities of GSM and Code Division Multiple Access (CDMA - a radio channel access method) units at the same time. Alternatively, a non-GSM-based system can be applied for disaster sites which are not covered by GSM network (e.g. remote refugee camps) or are temporarily without GSM coverage due to disastrous natural event. The eSOS communication arrangement is shown in Figure 3-8.

Figure 3-7 Solar panel mounted at the top of the eSOS toilet at optimal angle for maximum efficiency (Photo: D. Brdjanovic)

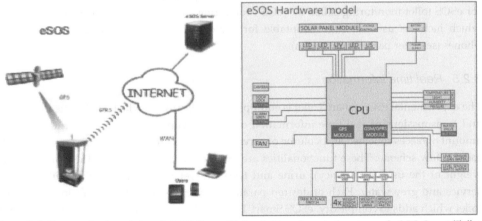

Figure 3-8 Communication chain of eSOS Smart Toilet (left) and control system of eSOS Smart Toilet (right) (SYSTECH.ba)

The GPS module and GSM/GPRS module make part of the eSOS central processing unit (CPU) which is electronic 'brain' of the toilet (Figure 3-9). All electronic components are connected to CPU. Wiring is hidden and protected against the atmospheric and human interference. The CPU, located safely inside at the top of the toilet structure, is tailor-made and produced by SYSTECH.ba, a member of eSOS Smart Toilet consortium.

Figure 3-9 Tailor-made electronic components of eSOS CPU (Photo: A. Muratbegovic)

The GSM antenna is located outside the toilet for better connectivity (Figure 3-10). Via GPS system each toilet is connected to a satellite so the geographic positioning of each unit can be determined exactly even if the toilet is relocated. Via PSM/GPRS signal the toilet is connected to a 'cloud' via internet, and dedicated eSOS server receives and stores all required information from the each eSOS toilet unit. Requested data and information are available for registered users with appropriate access rights for further handling and processing. A dedicated software for eSOS toilet monitoring called as eSOS Monitoring™ was developed by SYSTECH Bosnia which includes user interface adaptable for personal computers (PCs), tablets and smart phones (see later part of this section).

3.2.5 Real time information

The experimental eSOS Smart Toilet is equipped with electronics which enable measurements and transmissions of real time information on toilet occupancy, volume of urine collected, amount of faeces collected and volume of service water and grey water (see Figure 3-11 for the toilet's flow scheme. These functionalities are grouped as they all rely on the measuring the weight of the user (occupancy), urine and faeces excreted (individual or cumulative) and service and grey water. Each of desired parameters is measured at different location of the toilet which adds to complexity. eSOS Smart Toilet contains four storage components, namely (i) the storage of service water replenish-able by (combination of) rooftop rainwater harvesting, and manual filling by water vendor trucks or by piped water supply, all subject to availability, (ii) grey water collection tank which collects water from hand-washing facility attached to the back of the toilet and also water used for cleaning the interior of the toilet, (iii) urine collection tank which collects urine from users via urine diversion toilet (or squatting pan), and (iv) faces collection tank which collects faeces from user and from the anal cleansing where applicable (explained in latter sections).

Figure 3-10 GSM antenna at the top of the experimental eSOS Smart Toilet (Photo: D. Brdjanovic)

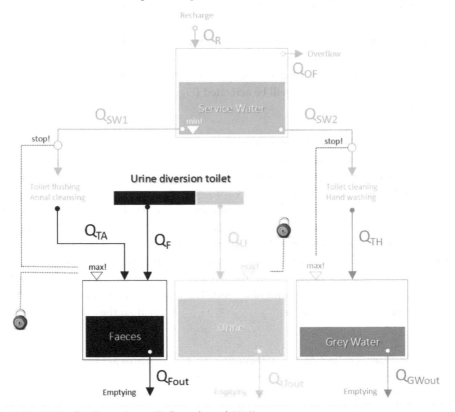

Figure 3-11 eSOS toilet flow scheme (Brdjanovic *et al.* 2013)

a) Occupancy scheme - occupancy could be observed by several methods (e.g. sensors) but it was decided to use weight cells as the information of the user's weight can be useful in data processing. In order to weight the user the arrangements of 4 sensors was established in the way that the floor of the toilet is rested on the 4 load cells where the weight of the floor structure (including the toilet bowl, if applicable) is deducted from the total weight when the person enters the toilet (Figure 3-12).

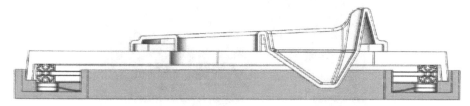

Figure 3-12 Squatting pan with load sensors in the scenario with squatting pan (identical arrangement applies in the case of pedestal) (Source: Flex/design)

Activation of these measurements also gives a signal of occupancy, the toilet is occupied until the moment that the person leaves the toilet and the weight measurement is back to 0. It is anticipated that while using the toilet, the toilet user would move around within the toilet cubicle of approximately 1 m^2. Thus in order to establish the accurate measurement of the user's weight, the averaging algorithm between the fours cells applies. Such degree of accuracy is not required to determine the occupancy, however it is needed when the accurate measurement of the weight of the person may be needed. As soon as the person steps in the toilet the occupancy exterior light will be activated (Figure 3-13) and the new use of the toilet will be registered.

b) Volume of urine collected - this measurement can also be obtained by introducing the urine level sensor, but it was decided to use weight cells as the level of accuracy and robustness drastically increase by their applications. As the high accuracy of urine accumulation (measurement) was required for the research purposes, the experimental toilet was equipped with the set of high accuracy industrial weight sensors. In order to have accurate measurement of urine accumulation, the urine tank was separated from faces tank, both resting on the stainless steel plate of thickness that would allow minimum deformation under the weight of the full tank. The deformation was measured by the structural endurance test which resulted in acceptable deformation of base plate (dimension 70x38x0.5 cm). Each sensor each was exposed to a force of 1,000N at the end of the test. The weigh plate shows bending of 0,37mm at the tips. The maximum stress in the plate is 4 times lower than the yield strength (Figure 3-14).

Both urine and faeces collection tanks, including the weight plates are supported by the base plate (dimension 60x60x0.5 cm). Each sensor each was exposed to a force of 1,000N at the end of the test. The maximum deformation of the plate is 0,4mm at the end of the load cell. Maximum bending in the base plate is less than 0,1mm. The maximum stress is around 4 to 5 times lower than the yield strength (Figure 3-15).

The weight of urine is expresses in grams and converted to volume of urine (mL) in ratio 1:1.

Figure 3-13 Occupancy light activated in eSOS Smart Toilet immediately after the user steps in (Photo: J. Ćurko)

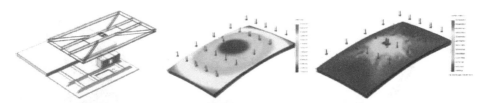

Figure 3-14 Assessment of deformation and stress on the base plate (Image: Flex/design)

Figure 3-15 Assessment of deformation and stress on the stainless steel plate supporting the weight plates and containers for urine (and faeces) collection (Image: Flex/design)

c) Amount of faeces collected - was determined in the same way as of urine (see Figure 3-16 for urine and faeces collection system). However, in this case the amount of water used for anal cleansing (also collected in this tank) was deducted thanks to the feature associated with anal cleansing button which activation can be programmed and registered (also at distance), so that the total amount of anal cleansing water per each toilet user (thus also the cumulative amount) can be accurately determined (e.g. for the research purpose).

Figure 3-16 Urine and faeces collection tanks (Photos: J. Ćurko)

d) Volume of service water and wash water - the volume is determined using very sensitive tailor-made (SYSTECH Bosnia) hydrostatic pressure weight sensors (Figure 3-17). Sensors are connected to CPU and the level (volume) in service water and wash water storage tanks can be determined with sufficient accuracy.

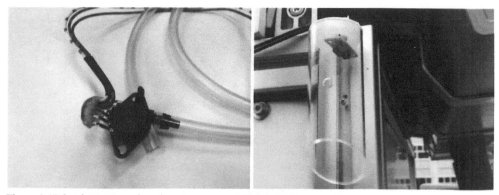

Figure 3-17 Service water and wash water sensors (Photos: D. Brdjanovic)

3.2.6   UV interior disinfection

The toilet is equipped with the UV lamp mounted at the top of the toilet (Figure 3-18). The UV lamp is set to be activated after each use of the toilet during the period that can be adjusted (during field testing duration of UV illumination was set to 3 minutes). In case that the next user enters the toilet before the de-activation time, the lamp will switch off automatically as soon as the new user is detected. The purpose of the UV toilet is to help maintain hygienic state of the toilet interior. The UV lamp functionality and efficiency was evaluated in laboratory and field conditions (Chapter 6).

Figure 3-18 Activated UV lamp in experimental eSOS Smart Toilet during testing at IHE Delft (Photos: J. Curko)

3.2.7 Nano-coated interior

To safeguard the interior surface of Smart Toilet from contamination, nano-coating was applied. This is especially important in regard to decrease affinity of toilet interface (squatting pan or pedestal) of retaining the traces of faeces at the surface. In this case a commercially available nano-coating was applied. The coating would create a hydrophobic surface where liquid would not precipitate once in contact to the surface, as demonstrated in Figure 3-19.

Figure 3-19 Comparison between tiles with nano-coating (left) and without nano-coating (right) (Photo: F. Zakaria)

Figure 3-20 Entry car reader feature of the toilet used by a child in Tacloban City, Philippines (Photo: F. Zakaria)

3.2.8 S.O.S. panic button

This feature is introduced inside the toilet to allow user to activate the sound alarm and in the same time send the signal to operator of the eSOS system who is supposed to notice the SOS alarm from the toilet on his/her PC, tablet or smart phone. Duration and intensity of the alarm siren noise is adjustable, also from the distance.

3.2.9 Odour control

The toilet under-structure is closed (only possible to open during the service intervals) and its ventilation is facilitated by the fan which is activated when needed. In default setup, the fan is activated at the moment the new user enters the toilet and run for definite period of time. It can also be possible to run the fan continuously, depending on the rate of charging the battery powered by a collar panel. At the end of the vent pipe, about a half meter above the roof, a replaceable granular activated carbon filter is installed to reduce or remove the odour leaving the toilet (Figure 3-21). It was also planned to design, manufacture and install the odour trap immediately under the toilet squatting pan or pedestal, however this feature was omitted and left to be solved during the design of eSOS final vision prototype. It is expected that this feature will further improve odour control in the eSOS Smart Toilet.

Figure 3-21 Granular activated filter for odour control in experimental eSOS Smart Toilet (Photo: J. Ćurko)

In addition to activated carbon filter at the ventilation, a urine odour trap was fastened to the urine outlet to urine tank (immediately after the toilet bowl). See the urine odour trap device at Figure 3-22. This was done anticipating the pungent urine smell.

Figure 3-22 Urine odour trap, inlet (left), silicon trap (right) (Photos: F. Zakaria)

3.2.10 Service water treatment unit

The experimental eSOS Smart Toilet is equipped with a 100 L storage reservoir for service water. The source of service water can be harvested rainwater from the roof, or from any other form of water supply. During the field testing period the reservoir (PVC tank) was coated in black to prevent sunlight penetrating the walls and stimulating deterioration of water quality. The service water is assumed to be of not drinking water quality and, therefore, eSOS team (Faculty of Food and Technology (PBF), University of Zagreb) designed and manufactured a mini water treatment system with granular activated carbon filter and UV disinfection lamp (Figure 3-23).

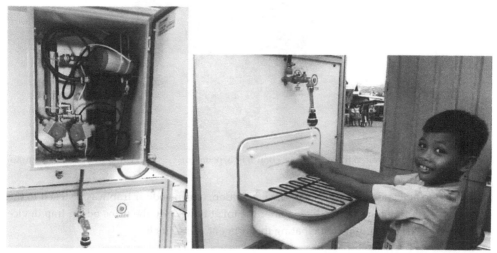

Figure 3-23 Mini unit for service water treatment (left) and a child using the hand washing facility (right)(Photos: F. Zakaria)

Cleaned water exit is split in two pipes, one leading to hand wash sink located under the treatment system at the back of the toilet, and one introduced inside the toilet for anal cleansing. Both lines are equipped with an electromagnetic automatic valve which opens for desired period of time (releasing known amount of water) when the user activate the push button either for hand washing or anal cleansing. Having the push button with limited flow per push prevents the tank from being quickly empty and water losses. The used wash water is collected in the so called grey water tank beneath the sink. The anal cleansing water, normally more bacteriologically polluted than the collected urine, is introduced in the faeces collection tank. The anal cleansing device is also to be used for interior cleaning of the toilet, if other cleaning means are not available. In this case the interior washing water will be entering a small opening at the bottom of the toilet floor and will be introduced to a grey water tank.

3.2.11 Software for monitoring and optimization

eSOS Monitor™ is a smart operations monitoring software which was developed by SYSTECH.ba and IHE Delft allowing each eSOS Smart Toilet to be located, monitored and serviced in an optimal fashion (Figure 3-24). This software is another innovative feature not applied before in the sanitation field. The software include hierarchical toilet organization, user-friendly marking with different icons or colours on the Google maps, filling level of tanks of urine, faeces, service water, grey water, reports of individual and cumulative toilet usage, statistical analysis of usage patterns and material flows, alerts/warnings using SMS, mail, intern messages etc., and many other interesting features.

Figure 3-24 Tablet version of the eSOS Monitor and its testing during the field application in Philippines (Photos: A. Muratbegovic)

The software is connected to all sensors in the toilet, further it governs all information flow that make functions of data recording, calculation and displaying on the internet connected web interface. The logic of the software is based on the different functions of the sensors and toilets features mechanisms which include flow scheme, occupancy scheme, and features functionalities scheme. The data is also reported in a web interface that display real-time information and retrievable cumulative data.

a) Flow scheme

Following the flow scheme as shown in Figure 3-11, the software logs each visit (making use of occupancy sensors), along with its associated water consumption (Q_A) , incoming flows to faeces tank ($Q_F + Q_{AF}$) and urine tank ($Q_U + Q_{AU}$). Having all data recorded in the sensors within logged time allow calculation to determine each visitors faecal sludge output (Q_F) and urine output (Q_U).

Cumulative visits data and its associated flows are recorded and retrievable using the web interface (see Figure 3-25). Further, details of flow by time during a visit which are stated as changes of tank's weight per 2-4 seconds, are also recorded and retrievable (see Figure 3-26)

In addition, as a results of flow measurements, the software could also report real time filling level of tanks as part of the software's web interface (see the top part of Figure 3-27)

Date/Time	Toilet	Duration	Urine	Dis. Urine	Faeces	Dis. Faeces	Discarded	Person	Gender
15/04/2015 06:31:23	TOL.NUM.1	04:02	0.02	0	0.16	0	0	37.87	MALE
15/04/2015 06:36:59	TOL.NUM.1	04:12	0.09	0.15	0.33	0.07	0.22	74.13	FEMALE
15/04/2015 07:54:47	TOL.NUM.1	01:26	0.21	0.02	0.03	0.33	0.35	48.64	FEMALE
15/04/2015 08:11:23	TOL.NUM.1	08:10	0.19	0.26	0.16	1.52	1.78	32.45	FEMALE
15/04/2015 08:52:34	TOL.NUM.1	07:28	0.14	0.34	0.28	0.1	0.44	34.45	MALE
15/04/2015 09:00:13	TOL.NUM.1	04:18	0.06	0	0.08	0	0	73.48	MALE
15/04/2015 12:31:04	TOL.NUM.1	09:07	0.07	0.06	0.17	0.95	1.01	35.88	FEMALE
15/04/2015 13:10:55	TOL.NUM.1	01:22	0	0	0	0	0	51.12	FEMALE
15/04/2015 13:22:48	TOL.NUM.1	05:40	0.4	0.22	0.02	0.16	0.37	33.16	MALE
15/04/2015 13:29:17	TOL.NUM.1	03:08	0.09	0.11	0.11	0.15	0.26	76.68	FEMALE
15/04/2015 13:57:02	TOL.NUM.1	05:01	0.04	0.04	0.02	0.15	0.18	43.44	MALE
15/04/2015 14:02:09	TOL.NUM.1	00:04	0	0	0	0	0	30.8	MALE
15/04/2015 15:14:37	TOL.NUM.1	05:19	0.18	0.17	0.02	0.01	0.18	80.22	MALE
15/04/2015 15:59:23	TOL.NUM.1	00:07	0	0	0	0	0	0	FEMALE
15/04/2015 16:00:32	TOL.NUM.1	20:05	0.01	0.09	0.24	0.92	1.01	32.01	MALE

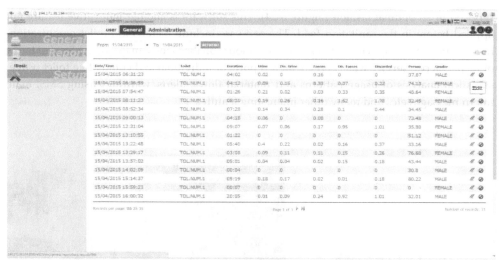

Figure 3-25 Cumulated logs of eSOS Smart Toilet in a day (Source: eSOS Monitor)

Date/Time	Urine	Faeces	Person	IWV
15/04/2015 06:31:23	1.42	9.53	0	false
15/04/2015 06:31:24	1.42	9.61	41.53	false
15/04/2015 06:31:25	1.43	9.61	40.1	false
15/04/2015 06:31:30	1.44	9.61	40.23	false
15/04/2015 06:31:33	1.44	9.61	40.43	false
15/04/2015 06:31:36	1.44	9.61	40.43	false
15/04/2015 06:31:47	1.44	9.62	40.1	false
15/04/2015 06:31:49	1.44	9.65	40.1	false
15/04/2015 06:31:51	1.44	9.66	40.1	false
15/04/2015 06:31:53	1.44	9.67	40.1	false
15/04/2015 06:31:55	1.44	9.68	40.1	false
15/04/2015 06:31:59	1.44	9.68	40.1	false
15/04/2015 06:32:16	1.44	9.68	40.1	false
15/04/2015 06:32:30	1.44	9.68	40.1	false
15/04/2015 06:32:43	1.44	9.69	40.1	false
15/04/2015 06:33:36	1.44	9.69	40.1	false
15/04/2015 06:35:25	1.44	9.69	40.1	false

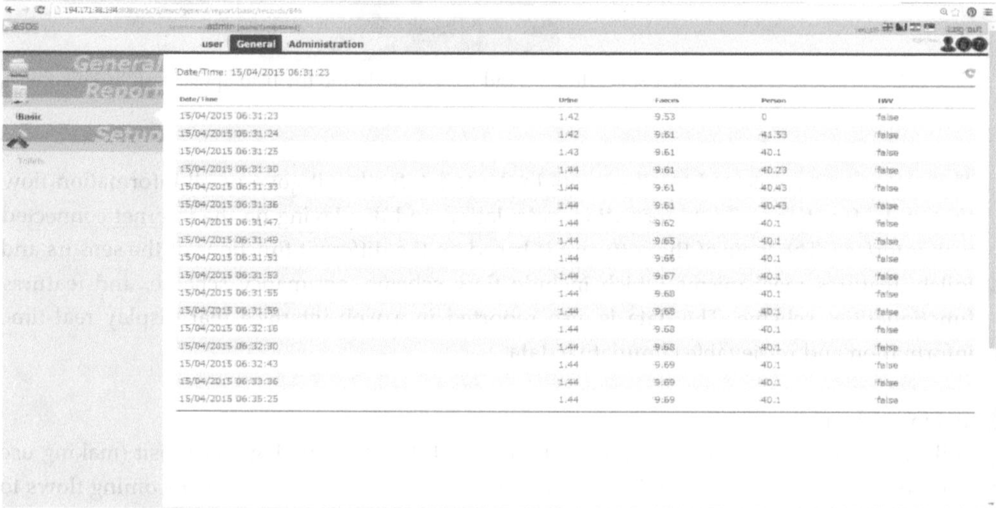

Figure 3-26 Individual log details (Source: eSOS Monitor)

b) Occupancy scheme

The toilet occupancy was detected using 6 weight sensors placed below the flooring platform (Figure 3-3). Once a person steps into the toilet cubicle, the sensors will send the data to the server where eSOS Monitor software will then mark the event and report that the toilet is occupied (Figure 3-27). At the same time, the door will be locked preventing other people to access the toilet despite having the key; the occupancy light (exterior light) will be on, and the interior light would also be on.

Figure 3-27 Web interface of eSOS Monitor web interface: when the toilet was occupied (left); when toilet is not occupied (right)

c) Features functionalities scheme

eSOS Monitor software also regulates timer function of the toilet's smart features i.e. service water withdrawal both for anal cleansing and handwashing that make use of solenoid valve, UV light, exhaust fan, mode switching from operation to service mode. In this regard, the software fixes how long water will flow each time the water button is pressed, how long the UV-C light will on to radiate the cubicle surface, and the interval of fan switching on. Other function to mention is locking down the toilet when one of the discharge tank is full or the toilet is occupied. In addition the software could also alert the coordination centre when the eSOS panic button is sounded.

All schemes could be compiled to a period of time of choices to present various trends, statistical data (Figure 3-28), various trends and usage pattern of the eSOS Smart Toilet (Figure 3-29 to Figure 3-32). The data can be aggregated to different streams i.e. urine, faeces, service water and grey water. In addition, the software make use of the GPS to locate the position of an eSOS Smart Toilet (Figure 3-33). This would be most useful in the up-scaled applications.

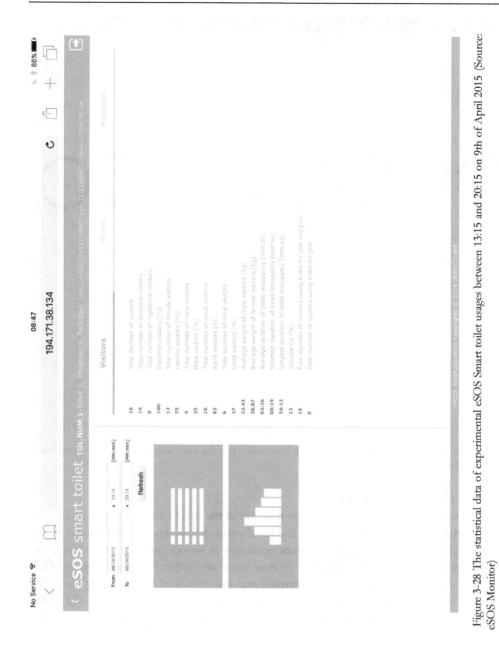

Figure 3-28 The statistical data of experimental eSOS Smart toilet usages between 13:15 and 20:15 on 9th of April 2015 (Source: eSOS Monitor)

<hr>

[1] The time shown at the x-axis of the chart suggests the Netherland's time, instead of the actual time in the Philippines

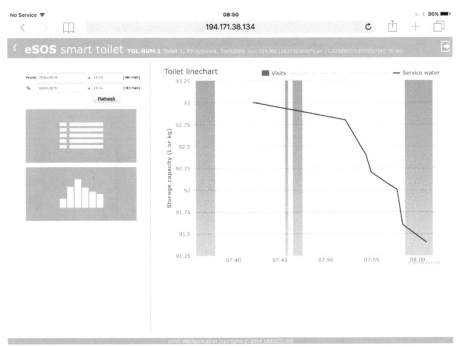

Figure 3-30 Water usages shown as decreasing trend in water supply tank between 13:15 to 14:15 am on 9th of April 2015 (Source: eSOS Monitor)

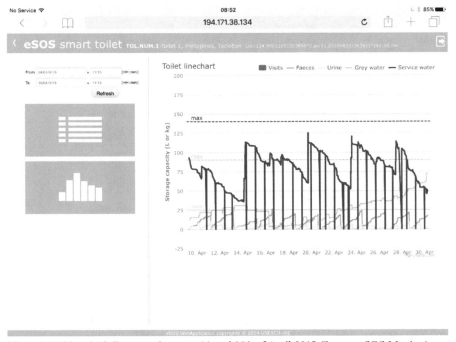

Figure 3-31 Trend of all streams between 9th and 30th of April 2015 (Source: eSOS Monitor)

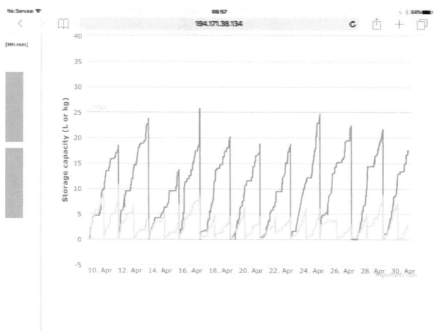

Figure 3-32 Trend of urine (yellow line) and faeces (orange line) streams between 9th and 30th of April 2015 (Source: eSOS Monitor)

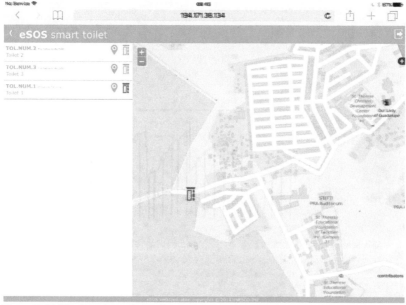

Figure 3-33 The location of experimental eSOS Smart Toilet in Abucay Bunkhouse, Tacloban City, the Philippines (Source: eSOS Monitor)

The eSOS Monitor makes a part of the larger eSOS software package which includes vehicle tracking and routing solution, and eSOS financial flow model. The eSOS Financial Flow Model Simulator that calculated cost elements of each sanitation chain has been developed (Figure 3-34). Validation of this model software is discussed in Chapter 9.

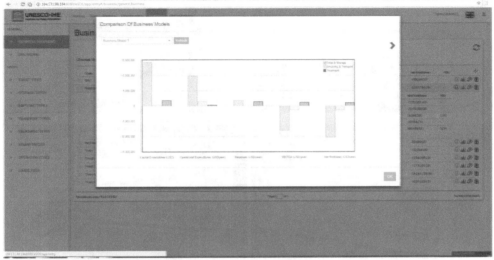

Figure 3-34 eSOS Monitor result graphs (Source: eSOS Monitor)

3.3 Deployment of eSOS experimental toilet for field testing

The experimental unit was first exhibited and tested (e.g. UV laboratory testing and other functionalities tests) at the IHE Delft garden from July 2014 to February 2015, when it was then dismantled to be shipped to the field-testing location at Tacloban City, the Philippines. The experimental prototype was not built as a light-weight and mobile unit as aimed for the final eSOS toilet product, therefore efforts were made to dissemble the toilet components and pack with safety for the cross continent deployment (Figure 3-35).

The toilet with other research related items were packed in 2 shipment crates which were weighted about 1300 kg in total, reached Tacloban City after 2 weeks by air freight from Schiphol Airport Amsterdam. It was flown to Cebu City, and then after custom clearance, it was transported by a truck which had to go across island by ferry to reach Tacloban City. The shipment was addressed to a warehouse of an international non-government organization (NGO) called Samaritan Purse. The NGO continued assisting transportation of the 2 shipment crates using forklift and truck (see Figure 3-36).

Upon arrival at the field testing site in at a temporary settlement in Abucay village at the outskirt of Tacloban City, the shipment craters were unpacked and the toilets parts were re-assembled and installed with the help of local community (see Figure 3-37). It only took a day and a half to reassemble the toilet structural parts. However, it took about a week to finish installation of ICT devices. Upon finishing the installation, the toilet was ready to undertake the field testing usage. The location of the toilet relative to the settlement can be observed in Figure 3-38.

Figure 3-35 Shipment Process in Delft – Top: Dissembled toilet structures were being loaded to the truck to further be packed by packing company; Bottom: Packing process by packing company into 2 shipment crates (Photos: F. Zakaria)

Figure 3-36 The Boxes arrived in Tacloban City, first stored in Samaritan-Purse's warehouse, then transported to the site with forklift and truck (with the help of SP) (Photos: F. Zakaria)

Figure 3-37 Assembling the toilet on-site with the help of local community (Photos: F. Zakaria)

3.4 eSOS Smart Toilet development overview

Following the field testing of experimental prototype, there are still phases to go before the toilet can be recognised as a viable product. The past development phases as well as future road map is presented in Figure 3-39.

Figure 3-38 Toilet location in the middle of the camp. This location was selected by the community representative (Photo: F. Zakaria)

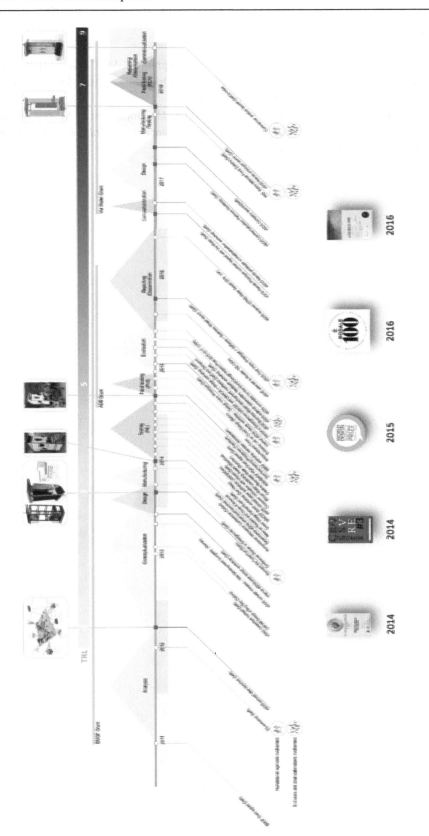

Figure 3-39 eSOS toilet development history

References

Brdjanovic D., Zakaria F., Mawioo P. M., Garcia H. A., Hooijmans C. M., Ćurko J., Thye Y. P. and Setiadi T. (2015). eSOS® – emergency Sanitation Operation System. *Journal of Water Sanitation and Hygiene for Development* **5**(1), 156.

Brdjanovic D., Zakaria F., Mawioo P. M., Thye Y. P., Garcia H., Hooijmans C. M. and Setiadi T. (2013). eSOS® innovative emergency sanitation concept. In: *3rd IWA Development Congress and Exhibition*, Nairobi, Kenya.

Evaluation of eSOS™ Smart Toilet

This chapter is adapted from:

Zakaria F., Ćurko J., Muratbegovic A., Garcia H., Hooijmans C. and Brdjanovic D. (2018) Evaluation of a smart toilet at an emergency camp. *International Journal of Disaster Risk Reduction (27), 512 – 523 (IF 1.603)*

Abstract:

An experimental prototype of the eSOS (emergency sanitation operation system) smart toilet was developed and tested at a transitional settlement of disaster affected people in the Philippines. The toilet was equipped with sensors and information communication technologies (ICT) for efficient operation in emergency setting. The field-testing aimed at evaluating the design of the toilet related to the user frequency/intensity, obtaining insight of usage patterns in a real-life situation, and testing the features and functionality of the toilet. The toilet gained data from nearly 700 users within a 7-weeks period. From the overall operational perspective, the toilet performed properly providing large, novel, and reliable information. It was evaluated at a maximum occupancy of up to 30 persons/day without queue, suggesting suitable application in transitional phase between short-term to long term emergency phases. Amongst the merits, the toilet saved up to 90% water consumptions compared to conventional toilets. The application of the eSOS toilet sensors and ICT allowed for a responsive maintenance resulting in optimum operation and minimum losses of users. The collected data gained insight in toilet usage pivotal to the design refinements of the toilet, as well as to improvements in terms of cost savings, better services and vision for sustainability.

Keywords: smart toilet; operation; usage; field-test; emergency

4.1 Introduction

People living in refugee camps are susceptible to displacement-associated diseases such as diarrhoea, which may cause high morbidity and mortality rates (Connolly *et al.* 2004; Waring & Brown 2005; Kouadio *et al.* 2011). Diarrheal diseases are transmitted predominantly through the faecal-oral route. Safe excreta handling, sufficient clean water supply, and proper hygiene practises are measures that need to be provided to intercept the transmission routes. That is, the sanitation provision at the emergency camps needs to take care of the entire sanitation service chain including containment/collection (i.e. toilet/latrine facilities), conveyance (sewerage for off-site system and desludging devices for onsite-sanitation system), treatment, and finally disposal or reuse.

Most of the emergency sanitation provisions opt for on-site sanitation systems. Servicing and maintaining on-site sanitation infrastructures have been proven to be challenging due to technical difficulties and under-investments (Parkinson & Quader 2008); particularly, at emergency settings. Both high population densities, as well as high flooding risks conditions, commonly observed at emergency settings, prevent digging toilet pits in emergency camps. Limited technical options suitable for the proper provision of sanitation under such challenging conditions (Zakaria *et al.* 2015) call for innovations (Bastable & Lamb 2012; Brown *et al.* 2012; Johannessen *et al.* 2012). Raised latrine systems using chemical or container toilets have been promoted to address the challenges (Morshed & Sobhan 2010; Bastable & Lamb 2012). They are (waterless) on-site sanitation systems with different servicing and maintenance mechanisms compared to other on-site sanitation systems which require the use of water e.g. a septic-tank system.

Several innovative container-based sanitation (CBS) toilets have been recently evaluated (Naranjo *et al.* 2010; Russel *et al.* 2015; Tilmans *et al.* 2015; Auerbach 2016). The results of these evaluations showed the need for strengthening the operation and maintenance (O&M) aspects of these toilets (e.g. the MobiSan™ and Uniloo™ toilets) in order to provide more reliable sanitation systems. The limited storage capacity of the containers demands a continued provision of tank emptying services to maintain a proper use of the toilets. That is, information related to the usage patterns of the toilets can be beneficial to better serve the toilets. Advancing on O&M aspects have been suggested to improve the performance of the recently developed and evaluated sanitation systems/toilets aiming at increasing the number of users and revenues, and reducing environmental, public health, and social issues, among others. Additionally, the extent at which the benefits provided by the sanitation system under emergency conditions reach the less privileged, including women and physically challenged citizens, is important information, as they appear to be poorly served by communal facilities in urban slums (Biran *et al.*, 2011).

In comparison to the provision of sanitation in conventional scenarios, the emergency context places additional requirements for the provision of proper toilets while maintaining their fundamental functionalities such as accessibility, safeness, and the provision of privacy

(Brdjanovic et al., 2015). All these necessities could be addressed by advancing on the O&M aspects.

Figure 4-1 shows the inadequate provision of latrines observed at an emergency context where a single toilet was shared by more than 20 persons even after the occurrence of the acute phase of the emergency (Cronin *et al*. 2008). Same trends have been reported in the case of long term informal settlements conditions (Wegelin-Schuringa & Kodo 1997). Improvements in the conventional monitoring and operation may ameliorate the condition of the toilets in case of a large number of toilet units, large area to be served, or an extremely high toilet usage frequency, as commonly found in camps where the usage pattern is unlikely to be consistent due to the dynamic of the displaced community. Monitoring of the use and status of conventional toilets, in order to develop a maintenance schedule, is normally done in person. An automatized monitoring and operational system would provide precise information regarding the toilet usage, and generate a responsive maintenance plan for the sanitation systems in such dynamic and crucial contexts.

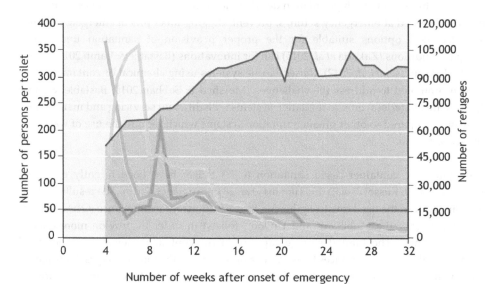

Number of weeks after onset of emergency

Figure 4-1. Average number of persons per toilet for 90,000 South Sudanese refugees in Uganda in camps. Adjumani camp (blue line); Arua camp (green line); and Kiryandongo camp (orange line). Violet shaded area shows the population fluctuation. In total 6,000 toilets were installed leading to 1:15 ratio i.e.15 persons per toilet (Modified from Murray (2015))

As response to such challenges, a novel (emergency) sanitation concept that uses an advanced monitoring system, eSOS™ (emergency Sanitation Operation System) was developed at IHE Delft (Brdjanovic *et al.* 2015). The eSOS Smart Toilet™, with its associated software eSOS Monitor™, is the key component of the eSOS concept. The eSOS smart toilet addresses the particular needs for the sanitation provision in emergency contexts including features such as easiness to be transported to the desired location, constructed with durable materials, minimum maintenance requirements, and no needs for excavation at the site. The toilet is also provided with self-cleaning capabilities, hand-washing facilities outside the toilet, interchangeable squatting pans or sitting toilet (for universal use according to local preference), separate faeces and urine collection tanks, smart lock system for protection and privacy, easy tank-emptying provisions, and a smart monitoring integrating ICT. The provision of the ICT features allows the eSOS smart toilet to optimise both the maintenance requirements, as well as the usage features of the toilet. Weight and level sensors placed at the faeces and urine collection tanks can alert the eSOS coordination centre whenever the tanks are almost full for collection. Additionally, the system locks the access to the toilet to avoid over-spilling of the collection tanks that may cause environmental hazards.

A prototype of the eSOS smart toilet was constructed and transported to Tacloban City in the Philippines. The prototype was evaluated by a typhoon-affected community located at a transitional settlement in Tacloban City. The objectives of this research were to: (1) evaluate the design of the toilet related to the user frequency/intensity; (2) obtain insight in the usage patterns of the toilet in a real-life situation by taking advantage of the ICT features; and (3) test the features and functionalities that provide the basis for improving the O&M aspects and continuous monitoring of the toilet. This study presents the main findings obtained at the field testing at the temporary settlement in the Philippines.

4.2 Research approach, materials and methods

4.2.1 Research design

An experimental prototype of the eSOS smart toilet was developed to evaluate its performance in an emergency camp. The eSOS smart toilet features and functions related to the three research objectives previously mentioned are listed in
Table 4-1.

Table 4-1 eSOS toilet features, functions, and relevance to research and testing objectives

Features	Intended functionality	Description	Relevance to objective (see Page 59)
Urine diversion (UD) pedestal	User interface	PVC unit diverting urine and faeces to separate storage tanks. Equipped with toilet seat, lid, and odour trap. Manufactured by Ecosave Netherlands	1, 2, 3
Service water supply reservoir	Service water storage	Operational volume 120 L, equipped with water level sensors mounted. Low level adjustable alarm feature.	1, 2
Wash water sink	Hand washing	Standard sink equipped with a water-saving tap and a water push button to control/optimize service water usage.	1
Faeces storage tank	Collection of faeces and water for anal cleansing	Operational volume 80 L, removable for manual emptying. Equipped with weight sensor. High level adjustable alarm feature.	1, 2, 3
Urine storage tank	Urine collection	Operational volume 80 L, removable for manual emptying. Equipped with weight sensor. High level adjustable alarm feature.	1, 2, 3

Table 4-1 Continued

Features	Intended functionality	Description	Relevance to objective (see Page 59)
Grey water reservoir	Hand wash water and interior cleaning water collection	Operational volume 120 L, equipped with water level sensors mounted. Low level adjustable alarm feature.	1, 2
Water buttons (2)	Hand wash, anal cleansing, and interior cleansing water supply	Push buttons for desired flow per push. The flow rate is set and regulated by solenoid valves.	2, 3
Bidet shower	Anal cleansing and interior cleaning	Standard bidet water-saving shower.	2, 3
Solar panel and battery set	Power generation	Dimensions 1.0 x 0.5 m, power, batteries 12V 7Ah sufficient for 7 days autonomy.	1,2,3
UV-C lamp	Interior surface disinfection	Mounted, switchable automatically, adjustable activity remotely (eSOS Monitor software)	Elaborated elsewhere (Zakaria *et al.* 2016)
Ventilation fan	Odour evacuation	Mounted in substructure, on-off adjustable.	1, 3
Granular activated carbon (GAC) filter ventilation	Odour control/treatment	Standard cartridge filter 100 g GAC, mounted.	1,3
Service water treatment unit	Treatment of service water	GAC filter and UV disinfection lamp.	2, 3
Smart lock system	Security/privacy	Activated by a electronic key to access toilet (in the form of a chip or card).	Elaborated elsewhere (Zakaria *et al.* 2017)
SOS panic button	Security	To be used in the case of emergency, it will sound an alarm to attract attention of the community member, as well as sending signals in the online monitoring system	Elaborated elsewhere (Zakaria *et al.* 2017)
Faeces and urine storage tanks weight sensors	Measuring and monitoring, locking the toilet.	Mounted, high accuracy, range 1 g to 150 kg. High level set point will lock the toilet to avoid over use.	1,2,3
Person's weight sensor	Measuring and monitoring, occupancy indication	Mounted, high accuracy 10 g to 150 kg.	1,2,3
Water level sensors (2)	Measuring and monitoring	Tailor made hydrostatic pressure electronic sensors for service water and grey water reservoirs.	1,2,3
GSM modem	Communication	Standard GSM module.	1,2,3
GPS tracker	Geographic positioning and tracking	Standard GPS module	1
Occupancy light	Indication of the occupancy	LED light, mounted, switch to red when the toilet is occupied.	2, 3, Elaborated elsewhere (Zakaria *et al.* 2017)
Light intensity sensor	Detection of light intensity	Standard sensor, adjustable manually.	Elaborated elsewhere (Zakaria *et al.* 2017)

Table 4-1 Continued

Features	Intended functionality	Description	Relevance to objective (see Page 59)
Thermometer	Measuring	To measure air temperature.	1
Control box	operation control and communication	Contain electronic boards and switches to allow for different modes of maintenance, operation, or reset. Linked with the eSOS Monitor software.	2,3
Roof gutter for rainwater harvesting	Service water harvesting	Standard PVC gutter.	1,3
Toilet structure	Stability, privacy, usage	Light-weight, semi-water-proof structure made from aluminium profiles and Dibond™ laminated sandwich panels.	
eSOS Monitor software	Operation and monitoring	Multi-feature dedicated eSOS Monitor Software.	1,2,3

The toilet was designed as a urine diversion (semi-dry) toilet with provision of water for hand washing, anal cleansing, and interior washing. Three collection and storage tanks (for grey water, urine, and faeces) were provided. The entire water and wastewater flow is shown in the schematic diagram in Figure 3-11. Each tank is equipped with sensors (weight or water level sensors) that can measure each tank content sending an alert when the tank is almost full and automatically locking the toilet.

4.2.2 Research location and community

The Abucay Bunkhouse was a temporary settlement located in Tacloban City, the Philippines, for families who had lost their home during the typhoon Yolanda that hit Tacloban City in December 2013. At the time this research was carried out (from February to August 2015), 199 families (813 individuals) were living at the settlement. The settlement consisted of 9 rows of buildings with between 10 and 27 household dwelling units per row. Shared sanitation facilities were provided in the camp consisting of two toilet blocks. Each toilet block was equipped with pour flush pedestal toilets and bathrooms. On average, three to four families were sharing one unit of toilets and bathrooms. The location of the toilet blocks is shown in Figure 4-2. The organization of the community was regulated by the municipality's social welfare office, whereby an official was appointed as a camp manager assisting the camp coordinator (a bunkhouse resident appointed by the community to represent the community).

4.2.3 Data collection and handling

The eSOS smart toilet was equipped with various sensors and other electronic equipment, connected through GSM/internet with a server for data storage. The data collected by the sensors was processed and made visible in real time or periodically (historic data) through a user interface called the eSOS Monitor™. The eSOS Monitor was also provided with built-in features allowing to change/adjust/control remotely the operation of the toilet and to assess the operational state of the toilet at any moment. In addition, some data was collected personally. Table 4-2 describes an overview of the data collection.

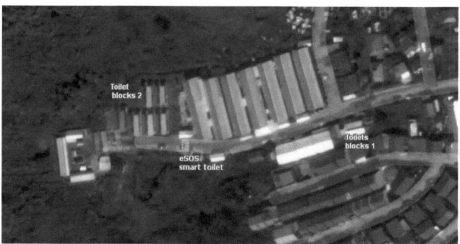

Figure 4-2. Situation map of Abucay Bunkhouse with location of eSOS smart toilet (photo: Google Earth)

Table 4-2. List of selected parameters of interest for the eSOS smart toilet field testing

Parameter	Indicator	Data source	Objective
Faeces tank emptying time and frequency	Dates of emptying, number of emptying events per week	EL	1,3
Urine tank emptying time and frequency	Dates of emptying, number of emptying events per week	EL	1,3
Cumulative daily usage	Number of visits per day	EL	2
Day-time and night-time usage	Number of visits at day and night per day (24 h)	EL	2
Daily male and female visitors	Number of visits by males and females per day	EL + ML	2
Daily adult and child visitors	Number of visits by adults and children per day	EL + ML	2
Daily defecation and/or urination	Number of visitors defecating and/or urinating per day	EL + ML	2
Duration of occupancy by males and females	minutes	EL	2
Duration of defecation and urination	minutes	EL	2
Amount of urine produced by male and female visitors	mL (measured as g) per visit and cumulative	EL	2
Amount of faeces produced by male and female visitors per visit	g per visit and cumulative	EL	2
Wash-water usage by male and female visitors, for urination and/or defecation, and by adult and child visitors.	mL (measured as g) per visit and cumulative	EL	1,3

EL: Electronic Logs; ML: Manual Logs

Daily logs from the data server (EL) were matched with the manual log book (ML) of the toilet usage to relate the collected information to the user's age, gender, and house location to increase the accuracy of the data acquisition. The most common sources of inaccuracy included the following: when only one toilet access key was used by multiple individuals of the same household, when household members had a similar body-weight, when young children were not able to fill in the information on the manual logging, and when users were occasionally using the toilet at the same time (mother with children), among others. By combining several set of data obtained from the different sensors, it was possible to get a detailed and frequently sampled (every 5 seconds) information at the level of individual user per each visit to the toilet.

The eSOS smart toilet was located at a convenient and safe location at the Abucay Bunkhouse camp (see Figure 4-2). After finishing the installation and preliminary functionality check, the toilet was available to the community 24 hours a day, 7 days a week, except when servicing and cleaning the tanks and the toilet.

The toilet was introduced and explained to the community, and the household representatives were asked about their willingness to use the new toilet. One access key to the toilet was provided for each willing-to-participate household. Each access key was numbered and linked to the corresponding household. As many as 93 keys were distributed to 91 households[2]. The household members older than 7 years of age[3] were registered by obtaining their names, gender, age, and body-weight with the approval of the corresponding individuals. The body-weight was necessary to link the toilet data to the gender and age of the user. In addition, the users were asked to fill-in a log book every time they used the eSOS toilet. The fill-in procedures consisted of writing down the access key number and the starting time of using the toilet. The field testing of the eSOS smart toilet was carried out for 49 consecutive days, from 13th March to 30th April of 2015. During that time, 662 valid[4] visits were registered. The toilet was cleaned every day with a customized cleaning procedure, including the use of an UV lamp for surface disinfection as in (Zakaria *et al.* 2016). The tanks were emptied once almost filled.

4.2.4 Data processing

Table 2 describes the evaluated parameters obtained electronically, manually, and by a combination of these two sources. Some parameters needed additional analysis. For example, to determine whether a person urinated or defecated while in the toilet, the observed patterns of measured faeces (Q_{Fout}) (as in Figure 3-11) and urine (Q_{Uout}) discharged to the collection tanks were calculated by processing the toilet usage data from 662 individuals. The eSOS Monitor software calculated the amount of anal-cleansing water based on the amount of service water drawn by each usage. The calculations showed the total amount of material

---

[2] Two households which were given more than one access key (e.g. in case of too many people per household).

[3] Children younger than 7 years old appeared to need assistance to enter and use the toilet.

[4] Visits during which a full set of data was obtained (about 25% of the total number of visits).

collected at the urine tank (Q_{Uout}) and at the faeces tank (Q_{Fout}). In addition, the amount of urine (Q_U) and faeces (Q_F) collected during one visit is calculated by subtracting the anal-cleansing water discharged to both tanks (Q_{AF} and Q_{AU}) to the total amount of material collected at both tanks (Q_{Uout}) and (Q_{Fout}).

4.3 Results

4.3.1 Determination of urination and defecation activities

Figure 4-3 and Figure 4-4 illustrate the observed different practices of urination and defecation respectively during the field testing. The value of different 'Q's (refer to Figure 2 for notation definition) are the total measured flow for each corresponding tank for that particular visit. Some assumptions were applied since there have been no references to distinguish urination and defecation. Defecation was assumed to produce larger amount of flow, taking relatively longer time and consuming more anal cleansing water. Applying these assumptions when observing individual usage data that was made into charts, it was found that defecation was likely to include sudden discharge of faeces to the faeces tank.

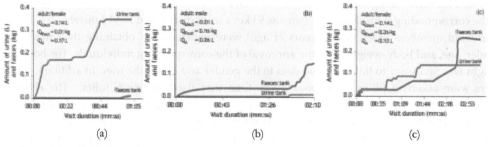

Figure 4-3 Individual practice in case of urination: (a) urination to urine tank only; (b) urination to faeces tank; (c) steady flow to urine tank with occasional discharge to faeces tank – end with cleansing water to faeces tank

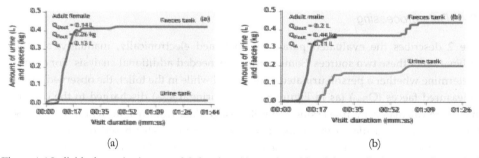

Figure 4-4 Individual practice in case of defecation; (a) practice with minimum discharge to urine tank; (b) practice with some discharge to urine tank

It was observed that defecation is most likely to include some urination. Figure 4-4a shows that even without any usage of cleansing water, some discharge to the urine tank was observed which indicate urination activity. It is more prominent as shown in Figure 4-4b, where a steady discharge into the urine tank was observed right after a sudden large discharge into the faeces tank (at 00:17 mm:ss); this trend can be explained by an urination event right after the defecation.

Having applied the assumptions and observations for defecations, urination was considered as such, as opposed to defecation, when the following conditions were observed: a dominant discharge into the urine tank, less or no water consumption for anal cleansing, a steady discharge flow within a short occupancy duration, and no sudden increase of weight in the faeces tank. Using this approach, it was possible to identify with sufficient confidence whether the user urinated or defecated; this information was valuable to quantify the amount of urine and faeces, to quantify the water consumption for anal cleansing, and to report the duration of these practices.

It was confirmed by observations that the urine tank received only urine and water except in very rare occasions when the stool size was so small that could escaped via the urine sieve. The faeces tank received all waste materials; faeces, wash-water, stool, and toilet paper.

Each visit was categorized either as urination (producing urine only) or defecation (producing urine and faeces). Both urination and defecation practices also produced anal-cleansing water stream (Q_A as described in Figure 3-11), which can be either discharged into the urine or faeces tank.

4.3.2 Operation of eSOS toilet during the field-testing

The field testing of eSOS smart toilet was carried out during 49 consecutive days with 662 valid visits of which 573 were by identified users. Figure 4-5 shows the number of visits per day, as well as the amount of urine and faeces collected in the storage tanks during the evaluated period.

At the early beginning of the evaluation, the toilet received a large number of first-time visitors. This number of visits decreased in the following days because of urine odour issues observed in the toilet cubicle due to malfunctioning of the urine odour trap. Therefore, starting on day 6 the urine tank was emptied each day, although the amount of urine did not exceed the 25 L maximum emptying threshold. This action resulted in a subsequent steady increase of the toilet usage to approximately 10 visits per day until day 19. A second drop in the number of visits to the toilet was noticed during the Easter holidays (*day 22 to 24*).

Shortly after, and for a period of approximately two weeks, the number of visits increased up to an average of 20 visits per day. The third drop on *day 37* was caused by the malfunctioning of the occupancy sensor; the night-time visits on that day were not recorded. Therefore, for the calculation of average users after *day 24*, the data from *day 37* was excluded. Further on, until the end of the evaluation period, the observed average number of visits reached 19 visits per day.

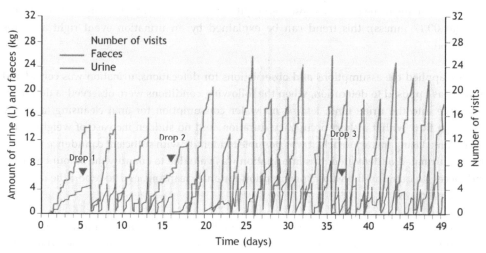

Figure 4-5. Number of visits to eSOS smart toilet including urine and faeces collection during the evaluated period. Drop 1, 2 and 3 indicate the time a sudden decrease in number of visits occurred.

The faeces tank was emptied every 3 to 5 days. When the toilet usage increased, the emptying frequency for the faeces tank became every 2 days on average (after *day 26*). In total, the faeces and urine tanks were emptied 19 and 43 times, respectively, during the evaluated period. The emptying criterion for the urine tank was the odour occurrence (as it happened that the urine odour trap was malfunctioning, so it was decided to empty the urine tank daily to minimise odour); on the other hand, the emptying criteria for the faeces tank was the maximum weight that one person could carry with ease. The aim of this field evaluation was not to operate the system at its maximum storage capacity, but to evaluate the usage performance of the toilet. The operation with extreme loads was evaluated during the commissioning period at the facilities IHE Delft, the Netherlands using surrogate materials. The maximum amount collected corresponds to 10 L of urine and 25 kg of faeces, with average values of 6 L and 20 kg, respectively.

4.3.3 Quantification of generated waste stream (urine, FS, cleansing water) per visit by user groups

As can be seen in Figure 4-6, the average amount of urine excreted by a male and female visitor were 170 ± 134 mL and 178 ± 130 mL, respectively. No distinction was made between adult and child visitors. Subsequently, the amount of generated faecal sludge was calculated, see Figure 4-7. The analysis revealed that the average amount of faeces excreted per visit by male and female visitor were 356 ± 250 g and 350 ± 240 g, respectively.

The recorded water consumption per visitor was also calculated in relation to gender, age (adult or child), and activity (urination or defecation), and it is shown in Figure 4-8. Males and females used approximately the same volume of water; comparatively, more water was used when defecating. Children used nearly 30% more water than adults. The average water consumption varied between 0.12 and 0.50 L per visit.

4.3.4 Usage patterns of the eSOS toilet

The information gathered both by the sensors and manually allows data disaggregation into defined categories to elucidate the eSOS smart toilet usage patterns. Figure 4-9 illustrates the toilet occupancy during day time (06:00 to 18:00 hours) and night time (18:00 to 06:00 hours). During the evaluated period, the average number of day- and night-time visits were 8.2 and 5.3, respectively, with a maximum of 21 day-time visits (on Day 27) and 10 night-time visits (on *day 48*).

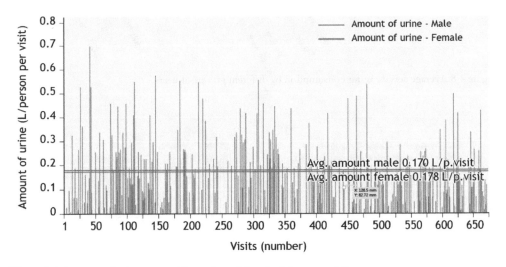

Figure 4-6. Amount of urine excreted by male and female visitors per visit.

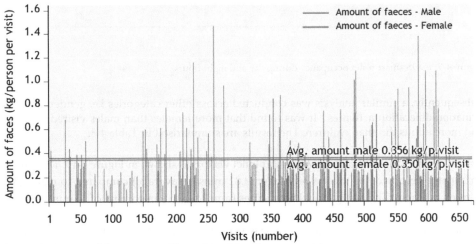

Figure 4-7. Amount of faeces excreted by male and female user per visit.

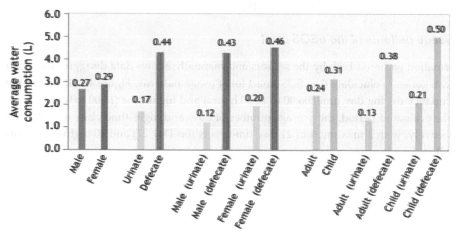

Figure 4-8 Average service water consumption by different groups of users

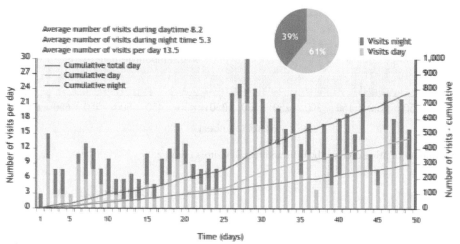

Figure 4-9. eSOS smart toilet occupancy during day and night hours.

Subsequently, a similar analysis was conducted across other categories i.e. gender, age, and urination-defecation activities. It was found that more females than males visited the toilet, and more adults did than children. The results are summarised in Table 4-3.

The occupancy duration of the eSOS smart toilet visitors is shown in Figure 4-10. The females on average occupied the toilet a bit longer (3.82 mins on average) compared to males (3.38 mins on average). The occupational time varies from 7 sec to a maximum of 20 min.

In comparison, the individual duration of stay in the toilet was related to the activity (urination or defecation - including urination) as presented in Figure 4-11. The average time needed for urination was on average 2.88 ± 2.5 min, while the average time needed for defecation and urination was on average 4.7 ± 3.5 min; that is, approximately 50% longer.

Table 4-3. Disaggregated toilet usage data into day-night-time usage, gender, age and activities categories

Parameters	Average daily visits	% of total users	% of total users (excluding unidentified users)
Male	4.4	32.6	38.0
Female	7.3	54.1	62.0
Unidentified	1.8	13.3	
Adult (≧ 18 years old)	7.0	51.8	59.9
Child (< 18 years old)	4.7	34.7	40.1
Unidentified	1.8	13.4	
Urination	8.3	60.0	
Defecation	5.1	40.0	

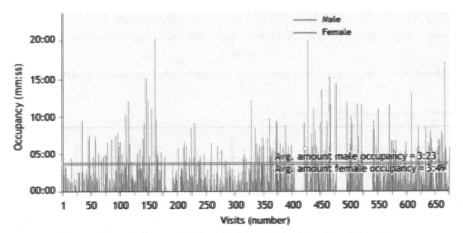

Figure 4-10. Duration of occupancy of the eSOS smart toilet by male and female users.

Further analysis of the data showed that the usage duration classifed per males urinating, males defecating, female urinating, and female defecating was 2.68, 4.12, 3.08, and 5.07 mins, respectively.

Out of the total reported day-time usage, only 34% corresponded to male usage, 53% to female usage, and 13% corresponded to unidentified or unregistered users. These figures are similar to the ones obtained for the night-time usage. Out of the total reported day-time usage, 58% corresponded to urination, while 42% for defecation. From the night-time usage observations more urination episodes were observed (64%), while only 36% was identified for defecation. Table 4-4 summarises these results.

The cross-category analysis generated additional insights, such as the increased usage for urination at night (58% day-time to 64% night-time). Males used the toilet equally for urination and defecation (50% urination, 50% defecation), while females used it more for urination (63% urination, 37% defecation). Contrary to the trends observed during the daytime for females, 55% of the daytime males used the toilet for defecation, while 64% of females used the toilet for urination day and night-time. Unidentified users are mostly those who visited the toilet for urination (67%).

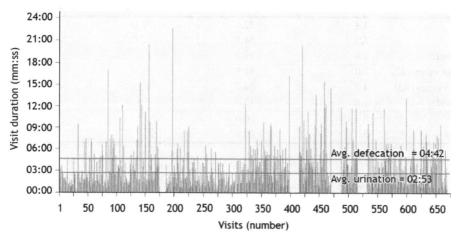

Figure 4-11. Duration of occupancy of eSOS smart toilet for visitors who only urinate or defecate and urinate.

Table 4-4 Summary of cross category analysis (%)

Category	Day	Night	Male	Female	Non-registered m/f	Adult	Child	Non-registered a/c	Urinate	Defecate
Day	NA	NA	34	53	13	52	35	13	58	42
Night	NA	NA	30	56	14	52	34	14	64	36
Male	64	36	NA	NA	NA	62	38	NA	50	50
Female	60	40	NA	NA	NA	58	42	NA	63	37
Non-registered m/f	59	41	NA	NA	NA	NA	NA	NA	67	33
Adult	61	39	40	60	NA	NA	NA	NA	57	43
Child	62	38	35	65	NA	NA	NA	NA	61	39
Non-registered a/c	59	41	NA	NA	NA	NA	NA	NA	67	33
Urinate	58	42	28	58	15	49	36	15	NA	NA
Defecate	64	36	40	49	11	55	34	11	NA	NA
Day-time male									45	55
Day-time female									64	36
Night-time male									60	40
Night-time female									64	36

4.3.5 Evaluation of eSOS toilet features and functionalities

From the overall operational perspective, the eSOS smart toilet functioned properly providing a large amount of novel and reliable data. Some features of the toilet were temporarily out of service; however, the issues were mitigated on the spot and valuable feedback was obtained for further toilet developments. All the 'smart' components of the toilet functioned as planned. Some issues were found after a few days of usage such as a dis-functioning urine odour trap at the UD pedestal. A few other issues developed later, e.g. the occupancy sensor (person's weight sensor) gradually lost sensitivity due to moisture and infestation of insects. Some actions taken at the site, as well as recommendations for improving the toilet design are as follows:

- The faeces tank was durable and showed no leaks, but it was transparent. The tank content was made invisible using duct-tape. The emptying of the tank was difficult to handle by one person because of the excessive weight of the tank. The tank emptying mechanism needs to be supported with additional features on the tank (e.g. tank with wheels) or an extra tool (trolley) needs to be added, so the tank can still be handled by one person; particularly, if the tank capacity is to be expanded.
- Smart lock system – access keys worked fine. However, the door lock needed periodical maintenance to ensure the door always locks properly.
- Person's weight sensor needs to be water-proof by sealing of the device.

4.4 Discussion

The discussion section is structured as follows: (i) design requirements considering the usage pattern data; this section includes both analysis of the toilet capacity in relation to the emergency phase, as well as the analysis of the tanks retention capacity in relation to the emptying frequency; (ii) usage pattern analysis by category; and (iii) O&M and monitoring.

The results of the field testing are valid for the chosen study location, time, and evaluated conditions. As the present study only involved one experimental eSOS toilet in the vicinity of other pre-existing toilets, the application in any different setting, for example a scaled-up field testing with many units of eSOS toilet where no other pre-existing toilets are present, might yield different results.

4.4.1 Design requirements in relation to usage pattern data

During the field testing, the experimental eSOS toilet received up to 30 visits per day. Because each visit in most cases was by a different person, it can be assumed that the eSOS toilet was used by 30 persons a day. According to SPHERE (Project 2011) and UNHCR standards[5], this

---

[5] Sphere and UNHCR standard for latrine provision lined that one latrine should be provided to every 50 persons in the short-term emergency and every 20 persons in the long term emergency phase.

usage corresponds to a transitional phase between short and long term phases commonly observed at emergency settings

The maximum serving capacity for the experimental eSOS toilet unit evaluated in this research was calculated. A maximum stay duration of 5.38 mins for urination and 8.20 mins for defecation was considered; in addition, a daily urination to defecation ratio of 60% to 40% was considered. Within a 24 hours period taking into account 8-hours of inactive and maintenance period, the toilet would have been capable of serving 105 urinating users and 45 defecating users; that is, a total of 150 users per day. In case of not considering urination and defecation details into account, and using an overall average visit duration of 3.62 mins, a total of 265 users per day can be estimated that may be able to use the toilet.

On the basis of this calculation, and considering the reality in the field where an emergency toilet may need to serve over 200 displaced persons per day at initial phase of emergency, the eSOS toilet has the potential to be used in the acute emergency phase.

In addition, the load accumulation rate for the faeces and urine tank can be calculated to find a suitable urine/faeces volumetric ratio for sizing the volume of the urine and faeces collection tanks. The calculations are presented in Table 4-5. Different usage/operation modes are included in this evaluation considering the actual usage (as in this research), waterless usage, and non-urine-diverting interfaces (that is, the use of only one tank with mixed urine and FS). This approach was considered useful for application in locations under different situations i.e. where users do not perform anal cleansing with water or where there are non-urine diverting toilet interface provisions. The integration of sensors and ICT into the eSOS toilet provided information for determining the flows of urine and faeces into the urine and faeces tanks. Combining this information with the previously analysed and reported data for urinating and defecating activities, it was possible to estimate the flow of urine and faeces into each tank; that is, the average flow per urination/defecation, and the urine and faeces flow percentage to each urine and faeces tanks (as presented in Table 4-5 Row 1 and 2). The exact mass of urine and faeces discharged into each tank was calculated and presented in Table 4-5 Row 6. Projecting for different usage/operation modes i.e. waterless usage and non-urine-diverting interfaces hence the use of only one tank (mixed urine-FS tank), usages without the use of cleansing water were grouped and tank flow data was averaged to be included for the tank volume ratio calculation. Similarly, the application of a mixed tank was also projected using the same data and calculation process. This approach is useful for application in locations under different situations i.e. where users do not clean with water or a non-urine diverting toilet interface.

Table 4-5 Calculation of tank volume ratio and corresponding retention time

No	Calculation formula/source	Actual — Urinate — Urine tank	Actual — Urinate — Faeces tank	Actual — Defecate — Urine tank	Actual — Defecate — Faeces tank	Waterless — Urinate — Urine tank	Waterless — Urinate — Faeces tank	Waterless — Defecate — Urine tank	Waterless — Defecate — Faeces tank	Combined tank with cleansing water — Urinate — Combi tank	Combined tank with cleansing water — Defecate — Combi tank
[1] Load per visit incl. anal cleansing water (average, kg)	Obtained data	0.35		0.91		0.15		0.58		0.35	0.91
[2] Flow (average, %)	Obtained data	52.0	48.0	25.6	74.4	60.9	39.1	17.4	82.6	100.0	100.0
[3] Urinate-defecate visitors ratio	Obtained data	0.6		0.4		0.6		0.4		0.6	0.4
[4] Tank capacity (L)	Actual	80	80	80	80	80	80	80	80	160	
[5] # of visits daily	Assumption	20				20				20	
[6] Load per visit per tank (kg)	[1]×[2]	0.18	0.17	0.23	0.67	0.09	0.06	0.10	0.48	0.35	0.91
[7] Mass volume (kg/L)	Estimated values	1	1	1.25	1.25	1	1	1.4	1.4	1	1.25
[8] Daily load volume to tank (L)	[6]×[5]×[3] /[7]	2.16	1.99	1.49	4.32	1.10	0.70	0.58	2.75	4.15	5.80
[9] Retention capacity of urine tank (d)	[4]/Σ[8] urine tank	22				48				16	
[10] Retention capacity – faeces tank (d)	[4]/Σ[8] faeces tank	13				23					
[11] Volume ratio	[9]/[9] ; [9]/[10]	1	1.7	1	1.7	1	2.1	1	2.1	N/A	N/A
[12] Simulated volume using [11] (L)	Correlated tank ratio to result in total volume of app. 160 L*	60	102	60	102	50	105	50	105	N/A	N/A
[13] Control retention capacity - urine tank (d)		16				30				N/A	
[14] Control retention capacity (faeces tank)-days		16				30				N/A	

* Current tanks volume is 80-L each to make up to total 160-L. This volume also represents the available space at the tank chamber – thus it is used as the basis for calculation.

Assuming 20 visits to the toilet per day, the daily load of each tank was calculated and converted to volume. The retention time, which is related to the emptying frequency, was obtained by dividing the designed tank volumes by the daily produced volume (Table 4-5 Rows 9 and 10). Finally, the urine/faeces tank ratio was obtained.

The usage of a waterless toilet reduced the load volume, as well as changed the flow segregation pattern to each tank. With almost the same volume, the emptying frequency for waterless usage could be 30 days, compared to 16 days for the usage with cleansing water.

4.4.2 Usage pattern analysis

There were more usages during the day than during the night. However, the fact that 39% of the usages took place at night suggests that the night-time related toilet features such as lighting and smart-lock system functioned properly during the night. This positive result is also supported considering that 40% of the female users visited the toilet at night and 38% of the child users also visited the toilet at night.

Further results were obtained from users interviews carried out as part of a user acceptance study conducted simultaneously to this research at the emergency camp (Zakaria *et al.* 2017). The users mentioned that one of the merits of eSOS toilet was the smart lock providing safety and privacy for the users. The toilet was regarded to be within reach, although a few people that lived at a building block further away admitted that they would have used the toilet more often if placed closer to them. In general, the results were positive compared to cases where toilets were not used. Reasons for not using the toilets include insufficient lighting and absence of locking features resulting in safety concerns e.g. threats of sexual assaults for female and child users, as reported to be the case for many for rural poor and urban slums dwellers. (Fisher 2008; Tilley *et al.* 2013; Kwiringira *et al.* 2014).

When segregating the usage by males and females, the results showed that the eSOS toilet was used more by females than males (despite there were about equal proportion of males to females in the bunkhouse). The male population at the camp did not necessarily need the privacy features of the toilet to urinate as it was observed that they were urinating in the open. In addition, adult males hesitated to change their urinating habit and be seated to urinate (as required by the eSOS toilet users guidance) since they are accustomed to urinate standing (user interviews). However, the data indicated that there were males who used the toilet to urinate, and that they usually do it during night (60% night-time male users urinated, compared to only 45% day-time male users that urinated). This might be because that they feel more secured urinating in the toilet at dark. It can be concluded that the provision of a male urinal would most likely not be effective at this study location, unless the urinal is in well-protected structure and well-lit at night.

Out of the entire identified usages, 60% were by adults and 40% were by children. An adult is defined as a user of 18 years of age or older. The toilet was regulated to be used by users older than or equal to 7 years old. Thus, despite this restriction, still having 40% of child users implies that the toilet was appealing to children.

It was observed that 60% of usages were for urination and 40% were for defecation. Considering that a person normally urinates 5-8 times a day (Schouw et al. 2002; Bael et al. 2007; Clare et al. 2009), and defecates 1-2 times a day (Rose et al. 2015), the number of defecation visits was relatively high. This may be attributed by the availability of other alternatives to urinate e.g. males that urinated in the open and other toilets. Also, most users did not stay at the bunkhouse all day.

On average females spent 3.82 minutes, and males spent 3.38 minutes in the toilet. The type of activity (urination or defecation) had more effect on the toilet occupancy time than gender. On average, users spent 2.88 mins to urinate and 4.70 mins to defecate. Nevertheless, although the difference between males and females was not significant, when the gender category was split up into urination or defecation, a more prominent difference was observed. Females appear to take nearly half a minute longer to urinate, and a minute longer to defecate than males. The time people spend in a toilet depends on many factors such as user's habitual routines, health condition, and many more, which might not be related to the toilet's functionalities.

Insignificant differences between males and females were also observed for the average generated amount of combined urine and faeces per toilet visit. Male users generated on average 170 mL of urine and 360 g of faecal sludge per person per toilet visit, while female users generated on average 180 mL urine and 350 g faecal sludge. Calculating the daily urine production per person that ranges between 600 – 2600 mL and average urination frequency of 6 times per day (Rose et al. 2015), then a person excretes between 100 – 430 mL urine every time. This study reports urine excretion at the lower side of that range; this may be attributed to less water consumption, hotter climate (people sweats more), or a combination of these factors. When comparing faeces production results obtained in this research with a compilation of wet faeces amounts reported by Franceys et al.(1992), the results fit into the suggested range of 209 – 520 g per person per day, assuming that the eSOS toilet users defecate once a day.

The results were reported after removing some abnormal usage data such as a case where an extremely high FS volume was discharged within a short period of usage (e.g. 3 kg of FS by a 12-year old boy who weighted 36 kg). Assuming on the fact that people keep a pot in their house for emergency-use, for children or for elderly during night, some users might have used the eSOS toilet to discharge their night soil, but this could not be confirmed.

Assuming that a person urinates twice and defecates once a day, using the same toilet, then this study shows that a person produces on average 360 g FS a day without anal-cleansing water. This result is within range with studies characterizing FS mass (Franceys et al. 1992; Rose et al. 2015). Norris (2000) showed an average sludge build-up rate of 0.07 L/persons/day for VIP latrines and 0.08 L/persons/day for septic tank systems, much lower than the findings of this study. However, this sludge build-up rate resulted from a combination of processes such as consolidation at the bottom of the pit or tank, leaching of soluble substances and evaporation) (Franceys et al. 1992).

The SPHERE standard recommends the use of a female to male ratio of 3:1 to calculate the required number of toilet cubicles. The findings of this research suggest a 3:2 ratio. This is considering that the proportion of toilet usage was 60% females and 40% males, and that the duration of females and males is not significantly different.

Females used only a bit more water i.e. 0.29 L compared to 0.27 L for males. The amount of water consumed depended more on the type of activity (urination or defecation). On average, users spend almost three-times the amount of water when they defecate (0.43 L) compared to when they urinate (0.17 L). When comparing the difference between adults and children, it revealed that children users tend to spend more water (0.31 L) compared to the amount that the adults used (0.24 L). Nevertheless, the overall water consumption of the eSOS toilet has proven that it uses significantly less water compared to the traditional pour flush toilets in use at the testing site. This is attributed to the combination of the mechanism (non-flushing) and the smart features (i.e. water button – solenoid valve) that reduce the water use. It was observed that people used minimal 1.2 L of water per toilet visit in their conventional toilet (Pean 2016). It can be concluded that the eSOS toilet reduced the water consumption for almost 90% compared to the current practice with pour-flush toilet. Considering that the water was scarce at the evaluated location, the water-saving feature of eSOS toilet was a valuable contribution for the community.

4.4.3 O&M and monitoring

The major disincentive of having container based sanitation system is the high maintenance cost due to frequent tank emptying. The sensors integrated with the online monitoring system allowed for a responsive maintenance resulting in a continuous use of the eSOS toilet throughout the evaluated period, except during the daily cleaning time. The faeces tank was emptied when it was at its maximum holding capacity (25 kg), and the emptying frequency was adjusted for the toilet usage (3-5 days at the beginning, 1-2 days later). Without the monitoring system, fixed periodical emptying would have been applied, and this would have caused either lack of efficiency or missing collections; thus, loss of toilet service capacity[6].

A simulation applying a fixed periodical emptying of 1-day, 2-days, 3-days, 4-days, and 5 days was made to support this argument. The analysis was conducted under steady state (same number of visitors generating steady daily FS of 5.18 kg) and dynamic state (fluctuating number of visitors and load as experienced in the field testing). When the cumulative FS production exceeds 25 kg, the toilet is considered closed from visitors resulting in loss of users for that day. The loss of service capacity and the number of emptying/maintenance performed for each simulated emptying period were calculated. Results are as presented in Table 4-6.

---

[6] Loss of service capacity implies to the loss of potential visitors that might have used the service of the toilet

Table 4-6 Simulation of applications of different fixed periodical maintenance compared to the eSOS-periodical maintenance at the field-testing period ranging between 2 to 3 days

		Emptying frequency (d)					
		1	2	This study	3	4	5
Number of maintenance events		48	24	18	16	12	9
Emptying efficiency vs. eSOS Smart Toilet's emptying		-167%	-33%	N/A	11%	33%	50%
Loss of service capacity	Dynamic	0%	4%	N/A	22%	35%	39%
	Steady	0%	0%	N/A	0%	0%	18%

When simulating a fixed periodical maintenance period of 4 days in comparison with the actual toilet usage, the faeces tank would have been emptied 12 times during the study period, instead of 18 times in practice, gaining a-33% maintenance cost efficiency. When simulated under dynamic state, this action would cause the loss of service capacity of 35% of the total toilet visits. Subsequently, at 3-day maintenance period, which was the closest to the actual operation trend of 18 times, gained 11% of maintenance efficiency, but still lost 22% service capacity. Increasing the emptying frequency to a 2-days maintenance period would need 24 times of maintenance (6 times more than eSOS operation), which means 33% loss of maintenance efficiency (33% more expenses), but still cause a loss of about 4% service capacity. When simulated under steady state, no loss of service capacity experienced until the emptying period is set to 5-days, in which there would be approximately 18% loss.

This simulation demonstrated that a fixed maintenance period would not result in optimum toilet maintenance for operation in an area where the toilet usage highly fluctuates, and that the monitoring system optimizes the maintenance efficiency at minimum maintenance expenses and service capacity loss.

A case of fixed maintenance period was demonstrated by the usage of the Freshloo Toilet by Sanergy in Kenya where daily collections were scheduled (Auerbach 2016). There was no report of missing collection in this case for the frequent collection schedule; however, the system would benefit from an optimized collection system using the responsive maintenance scenario. Particularly, because Sanergy operates in an up-scaled system, and it utilizes the collected FS to either produce fertilizer or biogas. An optimum collection would result in optimized operation expenses and optimum production of FS end products.

Pilot testing of CBS in Haiti which scheduled weekly collection reported occasional missing collections[7] that represent 0.5% of total FS removed by household service over the study period (Tilmans et al. 2015). Despite the small proportion that accidentally was released into the environment, it will represent dire risks should the event take place in an epidemic prone area,

---

[7] The CBS team operates by having weekly collection to each household subscribing to the CBS system services. The household users however have the access to remove the containment tank by themselves.

as it is frequently the case in most emergency settlements. The eSOS toilet was locked using the eSOS Monitor software, whenever the tanks reached its full capacity threshold. Nevertheless, the application of the eSOS toilet with its smart monitoring system has not been evaluated in an up-scaled application. Thus a 100% safe faecal sludge removal, although promising, cannot be concluded at this phase.

In summary, the eSOS Monitor software worked as expected during the field testing. The technology allowed for monitoring of all functionalities to optimise the operation and maintenance of the toilet, ensure safe FS management, and to validate the applicability and usefulness of toilet's functionalities in practice.

4.5 Conclusions and recommendations

The following general conclusions can be drawn: (a) the eSOS toilet features and its accompanying eSOS Monitor software was proven to work effectively during the field testing, (b) the ICT functionality allowed for continuous monitoring and remote adjustable operation of the toilet, (c) the smartness of the eSOS toilet was found useful to gain new insights in the design requirements of the toilet related to the frequency/intensity of use, usage patterns of the toilet in a real-life situation; and requirements for improved O&M and continuous monitoring of the toilet, and (d) the experimental toilet is currently at technology readiness level (TRL)[8] 5.

With regards to design requirements the following applies: (a) the toilet was evaluated by the occupancy of maximum 30 persons per day in a long term emergency phase. It was calculated that the eSOS toilet could serve more than 200 persons a day; thus, the toilet can also be applied in the short term/immediate emergency phase, and (b) urine to faeces tank volume ratio was calculated to be 1.0:1.7 for usage with anal-cleansing water, and 1.0:2.1 for waterless usage.

The most prominent outcomes related to usage patterns can be summarised as: (a) the eSOS toilet field testing generated data about toilet practice in detail that has never been obtained before, such as usage patterns by day-night time, by gender, by age group, by activity (i.e. it was possible to develop a methodology for automatic identification of urination and defecation), toilet occupancy time, faecal sludge and urine production per visit, and water consumption per visit, (b) the eSOS toilet was predominantly used during the daytime (61%), by adult users (60%), by female users (62%), mostly for urinating (60%), (c) male users tend to use the toilet to urinate at dark hours (60% of total male visits were at night), (d) male and females are not different with regards to the time spent in the toilet, amount of urine and faecal

---

[8] Using European Comission (EC)'s definition, TRL 5 is defined as technology where it has undertaken technology validated in relevant environment (industrially relevant environment in the case of key enabling technologies); TRL 7 – system prototype demonstration in operational environment and TRL 9 – actual system proven in operational environment (competitive manufacturing in the case of key enabling technologies; or in space) (European-Comission 2014)

sludge produced, and water consumption, and (e) differences in stay duration in toilet and water consumption depend on the type of activities i.e. urination or defecation.

Finally the most important finding of the study regarding the O&M practices (a) knowledge on the O&M under real usage allowed the performance of a proper evaluation aiming at achieving improvements in terms of cost savings, better services and vision for sustainability, (b) the eSOS toilet saves up to 90% of water compared to a conventional pour flush toilet, and (c) application of the eSOS toilet sensor and monitoring system allowed for a responsive maintenance. Application of such a responsive maintenance resulted in an optimum toilet usage efficiency by a minimum loss of users.

Based on the findings, recommendations are the following: (a) to continue with the prototype development to reach TRL 7 and ultimately TRL 9 using the feedback gained from the field testing and to test it for endurance and functionalities, (b) to develop a modular set-up (which has not been developed for the experimental prototype) and different types of eSOS Smart Toilet adjustable to socio-cultural requirements (urine diversion / non-urine diversion user interface, pedestal/squatting pans, anal wash water / waterless, etc.), (c) to develop the eSOS Smart Toilet Configurator (to allow for different custom-made configurations, and (d) to develop the eSOS Business Model Software (to calculate economic and financial feasibility to different eSOS applications globally).

References

Auerbach D. (2016). Sustainable Sanitation Provision in Urban Slums – The Sanergy Case Study. In: *Broken Pumps and Promises: Incentivizing Impact in Environmental Health* Thomas AE (ed.), Springer International Publishing, Cham, pp. 211-6.

Bael A. M., Lax H., Hirche H., Gäbel E., Winkler P., Hellström A.-L., Van Zon R., Janhsen E., Güntek S., Renson C., Van Gool J. D. and the European Bladder Dysfunction S. (2007). Self-reported urinary incontinence, voiding frequency, voided volume and pad-test results: variables in a prospective study in children. *BJU International* **100**(3), 651-6.

Bastable A. and Lamb J. (2012). Innovative designs and approaches in sanitation when responding to challenging and complex humanitarian contexts in urban areas. *Waterlines* **31**(1-2), 67-82.

Brdjanovic D., Zakaria F., Mawioo P. M., Garcia H. A., Hooijmans C. M., Ćurko J., Thye Y. P. and Setiadi T. (2015). eSOS® – emergency Sanitation Operation System. *Journal of Water Sanitation and Hygiene for Development* **5**(1), 156.

Brown J., Cavill S., Cumming O. and Jeandron A. (2012). Water, sanitation, and hygiene in emergencies: summary review and recommendations for further research. *Waterlines* **31**(1-2), 11-29.

Clare B. A., Conroy R. S. and Spelman K. (2009). The diuretic effect in human subjects of an extract of Taraxacum officinale folium over a single day. *The Journal of Alternative and Complementary Medicine* **15**(8), 929-34.

Connolly M. A., Gayer M., Ryan M. J., Salama P., Spiegel P. and Heymann D. L. (2004). Communicable diseases in complex emergencies: impact and challenges. *The Lancet* **364**(9449), 1974-83.

Cronin A. A., Shrestha D., Cornier N., Abdalla F., Ezard N. and Aramburu C. (2008). A review of water and sanitation provision in refugee camps in association with selected health and nutrition indicators – the need for integrated service provision. *Journal of Water and Health* **6**(1), 1-13.

European-Comission (2014). Technology readiness level (TRL). In, European Comission Decision C (2014) 4995 of 22 July 2014.

Fisher J. (2008). Women in water supply, sanitation and hygiene programmes.

Franceys R., Pickford J., Reed R., World Health O. and Who (1992). *A guide to the development of on-site sanitation*, Geneva.

Johannessen A., Patinet J., Carter W. and Lamb J. (2012). Sustainable sanitation for emergencies and reconstruction situations. In: *Factsheet of Working Group 8*, Sustainable Sanitation Alliance (SuSanA).

Kouadio I. K., Aljunid S., Kamigaki T., Hammad K. and Oshitani H. (2011). Infectious diseases following natural disasters: prevention and control measures. *Expert Review of Anti-infective Therapy* **10**(1), 95-104.

Kwiringira J., Atekyereza P., Niwagaba C. and Günther I. (2014). Gender variations in access, choice to use and cleaning of shared latrines; experiences from Kampala Slums, Uganda. *BMC Public Health* **14**(1), 1180.

Morshed G. and Sobhan A. (2010). The search for appropriate latrine solutions for flood-prone areas of Bangladesh. *Waterlines* **29**(3), 236-45.

Murray B. (2015). Average number of persons per toilet for 90,000 South Sudanese refugees in Uganda in camps of Adjumani, Arua, and Kiryandongo. In: *Presentation slides at WASH in Emergency Training by UNICEF-UNHCR* Delft.

Naranjo A., Castellano D., Zeeman G., Kraaijvanger H., Mels A. and Meulman B. (2010). The MobiSan approach: informal settlements of Cape Town, South Africa. *Water Science & Technology* **61**(12).

Norris G. A. (2000). *Sludge build-up in septic tanks, biological digesters, and pit latrines in South Africa*, Report 544/1/00, CSIR Building and Construction Technology, Pretoria.

Parkinson J. and Quader M. (2008). The challenge of servicing on-site sanitation in dense urban areas: Experiences from a pilot project in Dhaka. *Waterlines* **27**(2), 149-63.

Pean T. Y. (2016). *A framework to improve the product development process to achieve more effective sanitation response during emergencies*. Doctoral dissertation, Environmental engineering, Institut Teknologi Bandung, Bandung.

Project T. S. (2011). 4. Minimum Standards in Water Supply, Sanitation and Hygiene Promotion. In: *Humanitarian Charter and Minimum Standards in Humanitarian Response*, The Sphere Project, pp. 79-137.

Rose C., Parker A., Jefferson B. and Cartmell E. (2015). The Characterization of Feces and Urine: A Review of the Literature to Inform Advanced Treatment Technology. *Critical Reviews in Environmental Science and Technology* **45**(17), 1827-79.

Russel K., Tilmans S., Kramer S., Sklar R., Tillias D. and Davis J. (2015). User perceptions of and willingness to pay for household container-based sanitation services: experience from Cap Haitien, Haiti. *Environment and urbanization*, 0956247815596522.

Schouw N. L., Danteravanich S., Mosbaek H. and Tjell J. C. (2002). Composition of human excreta – a case study from Southern Thailand. *Science of The Total Environment* **286**(1–3), 155-66.

Tilley E., Bieri S. and Kohler P. (2013). Sanitation in developing countries: a review through a gender lens. *Journal of Water Sanitation and Hygiene for Development* **3**(3), 298-314.

Tilmans S., Russel K., Sklar R., Page L., Kramer S. and Davis J. (2015). Container-based sanitation: assessing costs and effectiveness of excreta management in Cap Haitien, Haiti. *Environment and urbanization* **27**(1), 89-104.

Waring S. C. and Brown B. J. (2005). The Threat of Communicable Diseases Following Natural Disasters: A Public Health Response. *Disaster Manag Response* **3**(2), 41-7.

Wegelin-Schuringa M. and Kodo T. (1997). Tenancy and sanitation provision in informal settlements in Nairobi: revisiting the public latrine option. *Environment and Urbanization* **9**(2), 181-90.

Zakaria F., Garcia H., Hooijmans C. and Brdjanovic D. (2015). Decision support system for the provision of emergency sanitation. *Science of The Total Environment* **512**, 645-58.

Zakaria F., Harelimana B., Ćurko J., van de Vossenberg J., Garcia H. A., Hooijmans C. M. and Brdjanovic D. (2016). Effectiveness of UV-C light irradiation on disinfection of an eSOS® smart toilet evaluated in a temporary settlement in the Philippines. *International Journal of Environmental Health Research* **26**(5-6), 536-53.

Zakaria F., Thye Y. P., Hooijmans C. M., Garcia H. A., Spiegel A. D. and Brdjanovic D. (2017). User acceptance of the eSOS® Smart Toilet in a temporary settlement in the Philippines. *Water Practice and Technology* **12**(4), 832.

5

Evaluation of water treatment and wastewater characterisation from eSOS Smart Toilet

Abstract

This chapter present the results of the laboratory analysis of the washing water and faecal sludge (FS), urine and grey water of an eSOS (emergency Sanitation Operation System) smart toilet that was used in an emergency settlement in the Philippines. A combination of tap/rain water was used for handwashing and anal cleansing, which was treated before use by a water treatment unit. Faecal sludge, urine and grey water were produced separately. Samples were collected at the study site at Tacloban City and transported on the same day to a laboratory at the closest major city (Cebu City). The efficiency of the toilet's water treatment unit was evaluated, while the FS, urine and grey water were evaluated to determine the necessary after-toilet management steps to ensure safe sanitation that is prerequisite in an emergency settlement. The results showed that the FS is of medium to high strength fresh FS, and the urine samples fits to fresh urine characteristics. Co-treatment of FS with wastewater in a treatment plant is not recommended, co-digestion with wasted sludge might be feasible. The urine and greywater require hygienization before being discharged to water bodies. The eSOS-Smart-Toilet's own treatment membrane unit effectively reduced the *E.coli* concentration, but it's capability to filter solids was not effective.

Keywords: faecal sludge, urine, characterisation, emergency sanitation.

5.1 Introduction

Amongst the challenges to avail proper FS management in emergency is the lack of knowledge about the composition of the waste streams that is essential to plan the handling process after collection. FS produced in emergencies is most likely 'fresh' or undigested, different from the sludge commonly collected from septic tanks that has been partially digested. Also, the type of toilet (wet vs dry toilets) determines the FS physical and chemical composition. In the case of diverting toilet, there is additional urine stream, whose disposal/reuse plan needs to be properly assessed. Different FS and urine characteristics might result in a different management plan.

An experimental prototype of the eSOS Smart Toilet was tested on operation and user friendliness in a transitional settlement in the Philippines. Characterization of FS, urine[9] and greywater discharges was conducted as part of the test. In addition, the water quality of influent and effluent of the toilet's own water treatment unit were also assessed to study the treatment-unit's efficiency.

The objective of the present paper is to assess the characterisation of FS, urine and greywater generated from the eSOS smart toilet tested in an emergency settlement (Tacloban City), as well as to evaluate the effectiveness of the water treatment unit. The findings will contribute to understanding the urine and FS characteristics in emergency settlement. The efficiency of the water treatment system will be used for the design improvements of the smart toilet.

5.2 Methodology

5.2.1 Experimental set-up

The experimental prototype of eSOS smart toilet has its own water supply (incorporating a rain water harvesting) as well as a water treatment system, connected to the shower head and wash hand basin tap. The water from the wash hand basin is stored in a grey-water tank. The toilet is a urine-diverting system separating faecal sludge and urine as can be seen in Figure 2-2. The toilet is integrated with ICT smart features (information – communication technologies), such as sensors to measure the flow to each tank.

5.2.2 Sampling

5.2.2.1 Water (fetched and rain water)

The water supply tank retained water from fetched water sourced from tap and nearby stream, as well as collecting rain water from the toilet roof. The water samples from the water supply tank were collected by opening a valve at the side bottom of the tank. The tank was supplied

---

[9] The eSOS Smart Toilet is a urine-diverting toilet, to generate FS and urine discharges

with manually fetched water sourced from uphill water stream and rainwater harvested from the toilet roof-top. Manually fetched water was added because the rainwater did not supply sufficient amount of water. The water was used for handwashing and anal cleansing.

5.2.2.2 Treated water

These were the water that was collected in service water tank and then flowed to the treatment unit (details of the water stream and treatment unit were discussed in Chapter 3). Treated water samples were collected from the hand-washing tap and put in a sample plastic flask.

5.2.2.3 Waste streams

Faecal sludge

The sampling took place around 10.00 – 12.00 in the morning at the scheduled sampling day, before discharging the tank content in the nearest communal septic tank. The faecal sludge in the tank was mixed evenly using a mortar mixer to achieve a homogeneous consistency. A predetermined volume and weight were scooped out of the tank into sampling flasks. One flask contained 650g of FS sample for chemical and physical parameters (i.e. Total Kejdahl Nitrogen (TKN), Ammonia-Nitrogen (NH_4^+-N), Chemical Oxygen Demand (COD), Total Phosphorus (TP), Volatile Solids (VS) and Total Solids (TS)) and another one contained 75g for *E. coli* measurement.

Urine

The urine tank was removed daily due to the bad odour observed in the toilet, and discharged in a nearby communal septic tank. The tank was shaken to mix the content, and approximately 100 mL was collected daily and accumulated over the same period of accumulation time as was applicable for the FS tank. At the FS sampling day, the accumulated urine sample was divided over 2 sampling flasks. One flask contained 750 mL for measuring the chemical parameters (TKN, TP, Potassium (K) and NH_4^+-N) and the other contained 100 mL for *E. coli* measurement.

Greywater

The greywater tank collected discharges from the toilet drain outlet at the floor and the hand-washing basin. The greywater tank was emptied once full. Greywater samples were collected at the same time as ther other waste streams samples. The tank was shaken to mix the content thoroughly before taking the sample from the outlet valve at the bottom of the tank. The samples in two sample flasks were then sent to laboratory be tested for parameters i.e. COD, TSS, TDS and *E. coli*.

5.2.3 Analytical methods

The following table shows the parameters measured and method used during the analysis. TKN, TP, K and Ammonia-N for urine and COD, BOD, TS, VS, TKN, TP and Ammonia were analysed using procedures outlined in the standard methods for the examination of water and wastewater (APHA *et al.* 2006). Each sample for each parameter was measured in triplication.

Table 5-1 Parameters and measurement methods

Parameter measured	Method used
Chemical Oxygen Demand (COD)	Closed Reflux, colorimetric
Biochemical Oxygen Demand (BOD)	Azide Modification (Dilution technique) (5 days at 20 ± 1°C)
Total Solids (TS)	Gravimetric
Volatile Solids (VS)	Gravimetric
Total Dissolved Solids (TDS)	Gravimetric
Total Kjeldahl Nitrogen (TKN)	Macro-Kjeldahl, Titrimetric
Total Phosphorus (TP)	Ascorbic Acid w/ Persulfate Digestion
Ammonia-N	Macro-Kjeldahl, Titrimetric (faecal matter) Phenate method, colorimetric (urine)
Potassium (K)	AAS, Flame Technique
Electrical conductivity	Electrical conductivity meter
Urine production per capita per usage	Toilet sensor
Faecal matter production per capita per usage	Toilet sensor
pH	pH meter (in situ)
Temperature	Thermometer (in situ)
Turbidity	Turbidity tube

5.3 Results and discussions

5.3.1 Water

The quality of rain/tap water and treated water were analysed, and then compared in order to calculate the treatment efficiency. The results were further compared to the Philippines water quality guidelines (Class B Recreational Water Class I[10] –(DENR-AO-2016-08 2016)) to check if the water quality meets the standards. This standard was used as the water treatment unit was not designed to produce drinking water quality standard. See Table 5-2 for details of the results and comparisons.

The treatment unit worked well enough to remove *E. coli* but it was less effective to remove solids. However, the solids quality still met the prerequisite standards for clean water. To cater a wide range of rain/tap water quality, it will be beneficial to add a stronger disinfection method to the treatment unit to ensure satisfactory treated water at all times.

The raw water contained some ammonia nitrogen which has probably got to the water from die off aquatic plants, land run off, animals, also potentially sewage leaked into the water source, as the raw water was fetched from a water pond as part of a small water stream nearby the study location. The treatment unit did not reduce any ammonia since there is no aeration

---

[10] Intended for primary recreational contact (bathing, swimming, etc.)

process in the treatment unit and short treatment time that there was not a possibility for oxidation to take place in the treatment unit.

Table 5-2 Water quality analysis of tap/rain water and treated water from eSOS toilet with comparison with Philippines clean water standard

Parameter	Raw water		Treated water		Removal efficiency (%)		Philippines clean water standard
	Min	Max	Min	Max	Min	Max	
Total suspended solids (mg/L)	<1	2.00	< 1	3.00	-50	ND*)	65
Total dissolved solids (mg/L)	123±14	153± 10	118 ±16	190 ±24	4	6	500
Total solids (mg/L)	123	155	118	194	4	5	
Ammonia-N (mg/L)	167	167	167	167	0	0	7 Nitrate as Nitrogen
E. coli (Colonies Forming Unit - CFU/100ml)	19	800	3	100	84.2	87.5	100 Most Probable Number - MPN/100 mL Faecal Coliform**
pH	7.62	7.85	7.41	8.02	-	-	6.5 – 8.5
EC (µs)	143.8	228	155.4	260	-	-	-
T (°C)	27.8	28.4	29.1	36.3	-	-	25 - 31

*) Not Detected

**) The standard only stated measurement unit in MPN instead of CFU, which can not be easily converted unless the same samples are analysed with corresponding method to get both MPN and CFU count. An attempt to convert MPN to CFU and vice versa concluded that E. coli in MPN is an order of magnitude greater than that in CFU, except in winter (Cho et al. 2010)

5.3.2 Wastewater

5.3.2.1 Faecal sludge

Laboratory tests and in-situ measurements of faecal sludge obtained from the eSOS toilet are presented in Table 5-3, including a comparison with literature.

The TS, TKN, NH_4^+, TP, pH and faecal matter production per capita per usage were found in the same range of referred literatures, but not necessarily fitting into one of the categories ('public toilet'/'septic tank'). VS, COD, BOD and E. coli were found to be much higher than the literature values. The high VS/TS ratio as calculated in this study i.e. 78-90%, instead of 45-73% in other studies, is related to a high organic content. This is attributed by the toilet being a urine-diverting toilet, less water usage for anal cleansing, hence increasing organic matter content in the faecal matter.

On average, the FS from the experimental eSOS Smart Toilet can be classified as medium to high strength FS following the classification as in Strauss et al.(1997) and Lopez-Vazquez et al. (2014). Exploring inexpensive treatment options, normally the co-treatment in a pre-existing wastewater treatment plant (commonly an activated sludge system) should be the first solution to be considered, as it is the easiest option should there is available facility nearby, particularly in the case of emergency situation where making use of surviving infrastructure may be the only viable option next to no-treatment at all. However, since the FS to be treated is of medium to high strength fresh FS with a high nitrogen concentration, subsequently requiring a lot of extra aeration (Dangol 2013) leading to high treatment costs and risks of overloading the treatment plant, co-treatment with wastewater should not be recommended as a treatment option in this case.

Table 5-3 Results of faecal sludge

Parameter measured	eSOS smart toilet		Public toilet	Septic tank	Reference
	Min	Max			
Total Solids (mg/L)	20,000 (500)*	47,000 (2,000)	52,500	12,000-35,000	Koné and Strauss (2004)
			30,000	22,000	NWSC (2008)
				34,106	USEPA and Agency) (1999)
VS/TS (%)	88	90	68	50-73	(Koné & Strauss 2004)
			65	45	(NWSC 2008)
Total Kjeldhal Nitrogen (mg/L)	1,200 ±30	3,900 ±900	34,000	1,000	(Katukiza et al. 2012)
Ammonia-N (mg-N/L)	79.5 ±0.6	1026 ±48	3,300	150-1,200	(Koné & Strauss 2004)
			2,000	400	(NWSC 2008)
			2,000 -5,000	<1000	(Heinss et al. 1998)
Total Phosphorus (mgP/L)	134 (15)	302 (13)	450	150	(NWSC 2008)
Chemical Oxygen Demand (mg/L)	330,000 ±71,000	43,000 ±16,000	49,000	1,200-7,800	(Koné & Strauss 2004)
			30,000	10,000	(NWSC 2008)
			20,000-50,000	<10,000	(Heinss et al. 1998)
Biochemical Oxygen Demand (mg/L)	146,339	146,339	7,600	840-2,600	(Koné & Strauss 2004)
E. coli (CFU/100ml)	1.0×10^{10}	$>2.0 \times 10^{12}$	1×10^5	1×10^5	Faecal coliform – NWSC (2008)
pH	6.25	7.63	1.5-12.6	-	(USEPA & Agency) 1999)
			6.55-9.34	-	(Kengne et al. 2011)

In absence of other viable treatment options, anaerobic co-treatment with waste sludge at a WWTP might be an option, if available. Anaerobic co-digestion is a viable treatment option when considering the FS' pH range that is between 6.25 and 7.23. (Strande *et al.* 2014). Thermal treatment of FS might be a viable treatment option as the water content is too high for drying the FS naturally. It was found that heating wet FS at a temperature higher than 55°C for two hours will lead to inactivation of *E.coli* (Turner 2002). The source of water in the faeces tank comes from anal cleaning by the users. Thermal treatment is fast but highly depending on power/energy availability. In the case of epidemic crisis, this treatment option seems the most effective, as it is safe and relatively quick.

5.3.2.2 Urine

The results of urine laboratory and in-situ analysis, as well as literature values, are presented in Table 5-4. It was found that the urine contains quite a wide range of high TP values, low values for potassium, ammonia and *E.coli*, and similar pH and conductivity values when compared to literature. It was considered diluted fresh urine considered that the wash water would have gotten to the urine tank, but when comparing with the fresh urine literature value, it is not fully consistent.

Table 5-4 Urine characterisation from eSOS toilet and their comparison with existing literatures

Parameter	Field results		Literature data						
	Min	Max	House-hold (1)	House-hold (3)	School (1)	Work-place (2)	Work-place (3)	Work-place (5)	Fresh urine (5)
Total Phosphorus (mg-P/l)	512 ±72	2993	210	313	200	76	154	540	800-2000
Potassium (mg/L)	62.9 ±0.1	111 ±6	875	1000	1150	770	3284	2200	2737
Ammonia-N (mg/L)	101 ±3	187 ±8	1691	3576	2499	1720	4347	8100	463
pH	8.73	8.98	9	9.1	8.9	9	9	9.1	6.2
E.coli (CFU/100ml)	2.0 x10^6	5.0 x10^8	<10CFU/ml (Schönning 2001)						
EC (µs)	8.71	17.9	1.1 to 33.9 mSn (Marickar 2010)						

(1) (Kirchmann & Pettersson 1994),

(2) (Udert *et al.* 2003),

(3) (Jönsson *et al.* 1997)

(5) (Geigy 1977)

(6) (Schouw *et al.* 2002)

(7) (Feachem *et al.* 1983)

The much lower potassium level in the urine sample compared to fresh urine level are possibly attributed by the difference in feeding habit, the amount of drinking water consume, physical activities, body size and environmental factors . People in the study location consume different type of diet that is probably of low potassium content, combined with the high humidity of the living environment to make people to sweat more, excreting some minerals with the sweat.

Evaluating the occurrence of high *E. coli* concentration, it is suspected that there must have been some cross contaminations from faecal matter, which is known to happen in source separated urine harvesting toilet (Höglund *et al.* 1998), as is the case with the eSOS toilet. This made options for direct reuse (for example for fertilizer) unfeasible. Even direct discharge would be unsafe. Storing urine helped in reducing the amount of pathogens (Vinnerås *et al.* 2003), which was evaluated for the purpose of hygienization of urine from urine-separated source with faecal contamination (Maurer *et al.* 2006). It takes minimal 6 month of storage to make urine completely safe to be used as fertilizer for any crop (Höglund *et al.* 2002), and the study evaluated urine with much lower *E. coli* concentration. Observing the much higher *E. coli* concentration in this study, additional disinfection treatment is required to meet minimum safety requirement for land disposal. (Maurer *et al.* 2006) assessed different types of treatment for urine hygienization (evaporation, acidification, micro- and nanofiltration, nitrification, electrodialysis and ozonation; concluding that electrodialysis and nanofiltration would potentially have strong hygienizational effects.

5.3.2.3 Grey water

The grey water analytical results together with effluent standards for the Philippines are presented in Table 5-5. It shows that most of the parameters met the discharge standards, except for *E. coli* and pH that was slightly too alkaline. Disinfection before discharging to a water body is needed to address the high *E. coli* concentration. Inexpensive disinfection methods to be considered include chlorination and solar disinfection.

Table 5-5 Comparison of grey water analytical results from eSOS toilet to Philippines effluent standards

Parameter	eSOS Smart Toilet		Effluent standards to inland waters (DENR-AO-34 1990)
	Min	Max	
Chemical oxygen demand (mg/L)	91 ±2	236	250
Total suspended solids (mg/L)	345 ±2	41 ±5	200
Total dissolved solids (mg/L)	231 ±5	234 ±24	2000
E. coli (CFU/100mL)	1.0×10^4	6.0×10^7	Max 500 MPN/100 mL if reuse to irrigate vegetable crop
pH	7.53	9.30	5.0 – 9.0
T (°C)	33.6	35.1	Max >3

5.4 Conclusion

Fetched/rain water and treated water

The treatment unit shows removal of the *E. coli* but was less effective to remove the solids and ammonia. Additional filtration and disinfection is recommended to enhance the treatment capacity of water treatment unit, ensuring a satisfactory output.

Faecal sludge, urine and greywater

The FS was classified as fresh FS of medium to high-strength on the basis of a comparison with literature values. The high *E. coli* content indicated possible pathogenic organism occurrences. The safest and fastest treatment might be thermal treatment to avoid an epidemic outbreak.

A high *E. coli* concentration was also found for the urine samples requiring treatment to sanitize the urine making it safe for disposal to water bodies or use as fertilizer. Storage alone would not be sufficient and would take a long time.

Greywater was found to be almost satisfactory for discharge to water bodies. The slightly higher *E. coli* content can easily be treated via quick disinfection such as chlorination or solar disinfection.

References

APHA, AWWA and WPCF (2006). *Standard methods for the examination of water and wastewater*. American Public Health Association, Washington, D.C.

Cho K. H., Han D., Park Y., Lee S. W., Cha S. M., Kang J.-H. and Kim J. H. (2010). Evaluation of the relationship between two different methods for enumeration fecal indicator bacteria: Colony-forming unit and most probable number. *Journal of Environmental Sciences* **22**(6), 846-50.

Dangol B. (2013). *Faecal sludge characterization and co-treatment with municipal wastewater : process and modeling considerations*. Unesco-IHE, Delft.

DENR-AO-34 (1990). In: Department of Environment and Natural Resources TP (ed.).

DENR-AO-2016-08 (2016). Water Quality Guidelines and General Effluent Standards of 2016. In: Department of Environment and Natural Resources TP (ed.).

Feachem R., Mara D. D. and Bradley D. J. (1983). *Sanitation and disease*. John Wiley & Sons Washington DC, USA.

Geigy C. (1977). Wissenschaftliche Tabellen Geigy, Teilband Körperflüssigkeiten. *Ed., Basel*.

Heinss U., Larmie S. A. and Strauss M. (1998). Solids Separation and Pond Systems. *For theTreatment of Faecal Sludges in the Tropics. Lessons Learnt and Recommendations for Preliminary Design*.

Höglund C., Stenström T. A. and Ashbolt N. (2002). Microbial risk assessment of source-separated urine used in agriculture. *Waste Management & Research* **20**(2), 150-61.

Höglund C., Stenström T. A., Jönsson H. and Sundin A. (1998). Evaluation of faecal contamination and microbial die-off in urine separating sewage systems. *Water Science and Technology* **38**(6), 17-25.

Jönsson H., Stenström T.-A., Svensson J. and Sundin A. (1997). Source separated urine-nutrient and heavy metal content, water saving and faecal contamination. *Water Science and Technology* **35**(9), 145-52.

Katukiza A., Ronteltap M., Niwagaba C., Foppen J., Kansiime F. and Lens P. (2012). Sustainable sanitation technology options for urban slums. *Biotechnology advances* **30**(5), 964-78.

Kengne I., Kengne E. S., Akoa A., Bemmo N., Dodane P.-H. and Koné D. (2011). Vertical-flow constructed wetlands as an emerging solution for faecal sludge dewatering in developing countries. *Journal of Water, Sanitation and Hygiene for Development* **1**(1), 13-9.

Kirchmann H. and Pettersson S. (1994). Human urine - Chemical composition and fertilizer use efficiency. *Fertilizer research* **40**(2), 149-54.

Koné D. and Strauss M. (2004). Low-cost options for treating faecal sludges (FS) in developing countries–Challenges and performance. In: *9th International IWA Specialist Group Conference on Wetlands Systems for Water Pollution Control and to the 6th International IWA Specialist Group Conference on Waste Stabilisation Ponds, Avignon, France.*

Lopez-Vazquez C. M., Dangol B., Hooijmans C. M. and Brdjanovic D. (2014). Co-treatment of faecal sludge in municipal wastewater treatment plants. In: *Faecal Sludge Management – Systems Approach Implementation and Operation.* Strande L and Brdjanovic D (eds), IWA Publishing, London, UK, pp. 177-98.

Marickar Y. F. (2010). Electrical conductivity and total dissolved solids in urine. *Urological research* **38**(4), 233-5.

Maurer M., Pronk W. and Larsen T. A. (2006). Treatment processes for source-separated urine. *Water Research* **40**(17), 3151-66.

NWSC N. W. a. S. C. K. S. P. K. (2008). Feasibility study for sanitation master in Kampala, Uganda.

Schönning C. (2001). Urine diversion–hygienic risks and microbial guidelines for reuse. *Solna, Sweden.* http://www. who. int/water_sanitation_health/wastewater/urineguidelines. pdf.

Schouw N. L., Danteravanich S., Mosbaek H. and Tjell J. C. (2002). Composition of human excreta – a case study from Southern Thailand. *Science of The Total Environment* **286**(1-3), 155-66.

Strande L., Ronteltrap M., Brdjanovic D., Bassan M. and Dodane P. H. (2014). *Faecal Sludge Management Systems Approach for implememntation and operation.* IWA Publishing.

Strauss M., Larmie S. A. and Heinss U. (1997). Treatment of sludges from on-site sanitation – low-cost options. *Water Science and Technology* **35**(6), 129-36.

Turner C. (2002). The thermal inactivation of E. coli in straw and pig manure. *Bioresource Technology* **84**(1), 57-61.

Udert K. M., Larsen T. A., Biebow M. and Gujer W. (2003). Urea hydrolysis and precipitation dynamics in a urine-collecting system. *Water Research* **37**(11), 2571-82.

USEPA and Agency) U. S. E. P. (1999). Decentralized Systems Technology Fact Sheet,Septage Treatment/Disposal. Document EPA 932-F-99-068. Office of Water, Washington D.C.

Vinnerås B., Björklund A. and Jönsson H. (2003). Thermal composting of faecal matter as treatment and possible disinfection method--laboratory-scale and pilot-scale studies. *Bioresource Technology* **88**(1), 47-54.

Effectiveness of UV-C light irradiation on disinfection of an eSOS™ Smart Toilet

This chapter is published as:

Zakaria F., Harelimana B., Ćurko J., van Devossenberg J., Garcia H., Hooijmans C. and Brdjanovic D. (2016) Effectiveness of UV-C light irradiation on disinfection of an eSOS™ Smart Toilet evaluated in a temporary settlement in the Philippines. *International Journal of Environmental Health Research (26) 5-6, 536-553 (IF 1.485)*

Abstract

Ultraviolet germicidal (short wavelength UV-C) light was studied as surface disinfectant in an eSOS™ (Emergency Sanitation Operation System) smart toilet to aid to the work of manual cleaning. The UV-C light was installed and regulated as a self-cleaning feature of the toilet, which automatically irradiate after each toilet use. Two experimental phases were conducted i.e. preparatory phase consists of tests under laboratory conditions and field testing phase. The laboratory UV test indicated that irradiation for 10 minutes with medium-low intensity of 0.15 – 0.4 W/m² could achieve 6.5 log-removal of *Escherichia coli*. Field testing of the toilet under real usage found that UV-C irradiation was capable to inactivate total coliform at toilet surfaces within 167 cm distance from the UV-C lamp (UV-C dose between 1.88 and 2.74 mW). UV-C irradiation is most effective with the support of effective manual cleaning. Application of UV-C for surface disinfection in emergency toilets could potentially reduce public health risks.

6.1 Introduction

People living in refugee camps are susceptible to displacement associated diseases such as diarrhoea, which may cause high morbidity and mortality rates (Connolly *et al.* 2004; Waring & Brown 2005; Kouadio *et al.* 2011). Diarrheal diseases are transmitted predominantly through the faecal–oral route. Safe excreta containment together with sufficient clean water supply and practise of proper hygiene, including hand-washing, are measures to intercept the transmission of diseases. Thus, the provision of safe sanitation is a life-saving response in the realm of emergencies.

Challenging environments which are often found in emergencies, such as densely populated areas, call for innovations in sanitation technical options (Bastable & Lamb 2012; Brown *et al.* 2012; Johannessen *et al.* 2012). The eSOS emergency Sanitation Operation System concept was presented as a promising alternative (Zakaria *et al.* 2015). The user interface, eSOS Smart Toilet, is a vital part of the eSOS concept. The toilet addresses the particular emergency requirements such as being easy to be transported, being made of durable materials, require minimum maintenance and do not require any excavation to install. The toilet is also advanced with unique features of having a smart monitoring and regulating system by means of integration of ICT (information and communications technology) and a sensors system, interchangeable squatting pans or sitting toilet (for universal use according to local preference), smart lock system for protection and privacy, easy tank-emptying design, and some ability for self-cleaning.

Being a place for defecation and urination, toilets are a potential source for spreading faecal-borne diseases. People might contract diseases by simply touching the surfaces of the toilet interior. When combined with poor personal hygiene, such as the absence of effective hand washing, infectious doses of pathogens may be transferred to the mouth (Rusin *et al.* 2002). The eSOS toilet is designed to be 'easy to wash and clean' (Zakaria *et al.* 2015). Having a clean (ideally pathogen free) environment inside the eSOS toilet reduces the transmission of diseases. This translates to both the selection of the proper toilet's interior material and the application of a self cleaning/disinfecting technology to aid on the common daily cleaning.

The importance of environmental disinfection has been extensively discussed in the context of hospital environment in relation to nosocomial infections (Cozad & Jones 2003). Nevertheless, studies have shown that bathrooms or toilets are recognised to be pathogens reservoir, whether in hospitals, schools or home environment (Gerba *et al.* 1975; Bloomfield & Scott 1997; Kagan *et al.* 2002; Barker & Jones 2005). Particularly, in overcrowding emergency settlements, where toilets are shared with many other users, the risk of contracting diseases in any of these pathways is much higher. With the more prominent infection risks present in emergency situations, emergency toilets disinfections are worth investing to aim for epidemic prevention in disaster-affected communities.

Toilets are commonly cleaned manually using water and cleaning chemicals, where cleaning action include brushing the toilet bowl, wiping toilet seat and wall surfaces, and mopping the floor. Depending on the degree of dirtiness, toilet cleanings are considered labour intensive.

Hence, manual cleaning of a toilet is done usually once or twice a day maximum. Chemicals are particularly used to help with visible stains removal and reducing smell which subsequently lessen the associated manual labours.

Manual cleaning is done after a period of multiple users. Therefore, this methodology cannot remove pathogens after each toilet use. To address this issue, automated or self-cleaning method has been searched to be used for public toilets. Research has been done on cleaning innovations that focus on the use of less or no chemicals, while still reducing the labour works. Those innovations ranges from cleaning tools such as micro-fibre cleaning cloth that is used without cleaning chemicals (Nilsen *et al.* 2002), laser cleanings (Gaspar *et al.* 2000) to self-cleaning surfaces in the form of choice of surface materials, coatings, etc.

Those cleaning innovations were explored to be incorporated as part of eSOS toilet surface cleaning regimes. The first self-cleaning technology candidate was the application of nano-coating. However, after preliminary laboratory testing, the nano-coating was found to wear off easily requiring frequent re-application. Thus, the idea of using nano-coating for self-cleaning technology in eSOS toilet was rejected.

The second method was ultraviolet (UV) germicidal light. UV is highly effective at controlling microbial growth and at achieving disinfection at most types of surfaces (Kowalski 2009). UV radiation in the wavelength range of 250 ±10 nm (UV-C) is lethal to most micro-organisms, i.e. bacteria, viruses, protozoa, mycelial fungi, yeasts and algae. The damage inflicted by UV-C involves specific target molecules; a dose in the range from 0.5 to 20 J/m^2 leads to lethality by directly altering microbial DNA through dimer formation (Bintsis *et al.* 2000). A low pressure germicidal UV-C lamp produces energy in the wavelength region of 254 nm. Schenk et al. (2011) revealed that 3 minutes of UV-C light irradiation at 253.7nm, that is a UV dose of 1.2 kJ/m^2, could achieve 7.2 log reductions of *E. coli* in liquid samples (i.e. irradiated cultured *E. coli* solution). This is in line with another study in the food industry, where a 0.67 – 1.13 log CFU *E. coli* reduction was achieved by UV-C irradiation on the surface of the cap of mushroom exposed to an UV-C dose of 0.45–3.15 kJ/m^2 (Guan *et al.* 2012).

In general, UV-C disinfection is widely used for drinking water treatment and air treatment. UV is also extensively applied for equipment sterilisation in the medical industry, and is fairly common in the food processing and packaging industry. It is less common, but has been explored for room disinfection, such as at hospital rooms (Andersen *et al.* 2006; Rutala *et al.* 2010; Stibich *et al.* 2011). These results were encouraging enough to support wider surface disinfection applications such as toilet disinfection.

Providing a toilet with UV-C surface disinfection feature adds the following advantages to eSOS Smart Toilet including: (i) easiness to clean; (ii) toilet disinfection after each use; and (iii) potentially less labour-intensive cleaning. However, the use of UV germicidal light exclusively for toilet surface disinfection has not been evaluated before. There is only limited knowledge about such application, thus its effectiveness needs to be determined.

This study aims at evaluating the effectiveness of using UV-C light irradiation on the disinfection of surfaces on the eSOS Smart Toilet under real usage of the toilet. The disinfection capacity of UV-C irradiation was first evaluated at the premises of IHE Delft using an experimental prototype of the eSOS Smart Toilet, followed by an evaluation using the prototype at a real emergency camp in Tacloban City in the Philippines.

6.2 Materials and methods

6.2.1 Research design

The research was divided into two phases. The first phase was a preparation phase in the IHE Delft campus the Netherlands. The second phase was the field research evaluation of the experimental eSOS Smart Toilet in Abucay Bunkhouse in Tacloban City, the Philippines.

The aim of the first phase was to relate the UV-C light intensity with the different locations in the experimental eSOS Smart Toilet. Furthermore, a direct effect of UV-C irradiation on *E. coli* deactivation was measured in the laboratory set-up, using a collimated beam and agar plates inoculated with a known amount of *E. coli*. Both procedures were necessary to determine and set the required dose of UV-C light for disinfection. Finally, it was necessary to develop and calibrate a method to measure the indicator bacteria concentration on the different surfaces of the toilet. The second phase during the field research assessed the UV-C light effectiveness and different cleaning methods under real toilet usage at the Abucay Bunkhouse in Tacloban City, the Philippines. The assessments on this second phase were evaluated using the detection of indicator bacteria (i.e. *E. coli* and total coliform) from sampled surfaces.

A prototype eSOS toilet serving as an experimental platform was developed for this research. The prototype exterior and interior outlook are shown in Figure 6-1. This experimental prototype was equipped with all features as aimed in the eSOS concept visions, such as urine diversion user interface; collection tanks for subsequent fast and easy desludging; external sink, weight sensor, emergency button, smart lock system, UV-C lamp and solar panel to power the toilet operation, and computer software. Users of experimental eSOS Smart Toilet are provided with water for anal cleansing and hand-washing through the toilet's own water supply system. The water can be used for a bidet shower. The water flow is regulated by pressing a 'water button' that is located next to the shower head. After pressing the water button, the water flows through the shower for 10 seconds. Like all other functions of the eSOS toilet, the water flow is controlled by the computer software.

Since the toilet is equipped with various sensors that work in unity with the toilet's equipment and mechanisms, the controlling computer program called eSOS Monitor (see Chapter 3) is used to integrate, track, record and monitor all functions. Real time toilet operation data and measured data by the sensors are also made accessible through internet connection to allow remote monitoring.

Figure 6-1 Experimental eSOS Smart Toilet at field test site in Tacloban, the Philippines (left); the interior (right) (Photos: F. Zakaria)

6.2.2 Experimental procedures and measurements

6.2.2.1 Phase 1: Preliminary research before field research at IHE Delft campus, the Netherlands

UV-C light characterisation. The experimental eSOS Smart Toilet was provided with a UV-C lamp (low pressure compact UV-C lamp, TUV 11W PL, Philips, Netherlands) that was situated at the indoor ceiling of the toilet. The UV-C light intensity was measured using a radiometer/photometer (Photometer ILT 1700, International Light Technologies 10 Technology Drive Peabody). The photometer was calibrated 1.5 years before the experiments and later compared with a more recently calibrated photometer for reading accuracy and was observed to provide satisfactory results (deviation of 1% reading in 2 decimals W/cm^2).

Light intensity measurements were conducted at the experimental toilet provided with the UV lamp at the ceiling. The measured sampling points at the toilet included the upper and lower walls, and parts of the toilet that most likely will be touched by users of the toilet (i.e. door handle, water button, shower head, S.O.S button and toilet seat). The photometer was placed in the sampling spot, and the reading was noted as the UV-C light intensity on that particular sampling spot. The linear distance from the spot to UV-C light was measured; any obstruction from the UV-light by a certain object was noted. The photometer was placed at different angles relative to the UV-C light to measure the UV-C light intensity at certain measuring spots. For example, at the water button measuring spot, situated at the toilet wall, the light intensity was measured placing the photometer at horizontal, vertical and perpendicular positions with respect to the source of UV-C light. While for the floor samples, whose surfaces were horizontally flat, the measurements were done placing the sensor on a vertical position with respect to the surface of the floor. The UV-C intensity was also measured directly under the UV light source to determine the maximum intensity. Figure 6-2 shows the sampling points where the light intensity was measured. Toilet seat was assessed under different conditions i.e. when the lid is down (Point 11) and up (Point 12) (Figure 6-2).

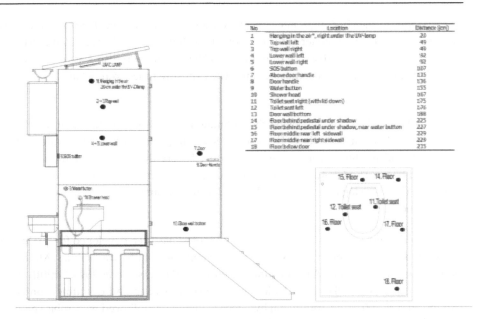

Figure 6-2 Schematic drawing of the experimental eSOS Smart Toilet with sampling points: side view (left); top view (right) (Image: F. Zakaria)

UV-C light irradiation test at the collimated beam apparatus and at the eSOS toilet. The UV-C light irradiation tests to assess the effectiveness in disinfecting *E. coli* were conducted both on a collimated beam (28W low pressure UV lamp, Berson UV Co., The Netherlands), and on the experimental eSOS Smart Toilet provided with a UV-C lamp (11W low pressure compact UV-C lamp, TUV 11W PL, Philips, The Netherlands) placed at the ceiling of the toilet. Both evaluations were carried out at the premises of IHE Delft, The Netherlands. The collimated beam was located at the laboratory facilities, while the experimental eSOS Smart Toilet was situated outside the building.

Laboratory cultured *E. coli* (non pathogenic *E. coli* ATCC 25922) was used as test organism. The initial concentrations of the cultured *E. coli* ranged from 0.9×10^9 to 1.3×10^9 CFU/mL. The cultured *E. coli* solution was diluted to pre-determined concentrations and spread on the standard agar (Chromocult™, Merck, UK) petri dishes (9 cm in diameter, corresponding to an area of 64 cm^2). Inoculated petri dishes were incubated for approximately 24-hours at 37°C. The purple and the red colonies that grow on the Chromocult were identified as *E. coli* and total coliform bacteria colonies, respectively. This part of the research only considers *E. coli*.

The efficiency of UV-C was observed by comparing the *E. coli* colonies count in UV-C exposed petri dishes with petri dishes that were not exposed to UV-C light (i.e. blank petri dish). *E. coli* log removals were calculated based on the results.

The collimated beam experiments were performed at a light intensity of 0.4 W/m^2. The intensity was set by adjusting the distance of the petri dish to the UV-C light. This intensity corresponded to the UV-C light intensity at the water button location at the experimental toilet measured with the photometer at the horizontal position. An *E. coli* inoculated petri dish was put at that point of fixed distance from the UV-C light. The petri dish was removed from the UV-C light after the planned exposure time. The exposure times were set at 1, 2, 3, 5, 7 and 10 minutes.

Tests similar as the UV-C test using the collimated beam were conducted at the eSOS toilet using the UV lamp at the ceiling. The selected sampling spots included the water button, the toilet seat, and the floor. The exposure times used were the same as for the test using the collimated beam. The UV-C light was turned on and off remotely using the toilet controlling program. The *E. coli* log removal was calculated using the same procedure as for the collimated beam experiment.

Standardised hygiene swab test. A surface cleanliness test has been developed for quality control in food industries as well as for hygiene requirements in health care facilities. Methods such as adenosine triphosphate (ATP) bioluminescence test (Davidson *et al.* 1999; Griffith *et al.* 2000; Aycicek *et al.* 2006), RODAC™ (Replicate Organism Detection and Counting) contact plates (Andersen *et al.* 2006) and traditional hygiene swabbing test followed by aerobic colony count (ACC) (Davidson *et al.* 1999; Griffith *et al.* 2000; Aycicek *et al.* 2006) are being used. The traditional hygiene swabbing test followed by ACC was selected for this research, because of its feasibility to be conducted in the field.

The challenge of using the swab test is that there is no universally accepted swabbing protocol (Moore & Griffith 2007). Subsequently, swabbing recovery rates were not known as they are specific to the experimental conditions. The swab recovery rates cannot be universally applied; therefore in-house standardisation is needed.

Swabbed samples were planned to be collected during the experimental eSOS Smart Toilet field testing research; therefore, the standardisation of the swab test and the assessment of the swab recovery rates were evaluated at a preliminary laboratory phase. Assessment of the swab test recovery rates were conducted for wet and dry surfaces at different surface materials corresponding to the materials used in the experimental toilet.

Realizing the limitations of equipment and materials in the field, the swab test method was designed to be simple in a way that it could be done without proper laboratory facility. Dry cotton swabs packed in individual sterile tubes were used. Two mL of physiological saline solution 0.9% NaCl in distilled water was added to the swab tube as the wetting solution. The surface sampling area was defined by a 9 cm diameter circle, i.e. the size of a petri dish. The swabbing was consistently done throughout the entire evaluation; that is, each side of the swab tips stroke about the same proportion of the sampling area. The swab was done at about 15 – 17 strokes each vertically and horizontally across the sampling area.

The three types of surfaces found on the experimental eSOS Smart Toilet were evaluated including steel (as observed in the shower head and door handle at the toilet), floor material (i.e. Trespa™ - a high pressure laminate of wood or paper fibre's bonded with a phenolic resin), and wall material (i.e. Dibond™- sandwich panels with a thin aluminium sheet on both sides laminated on a core of Polyethylene). Different conditions were also studied i.e. wet and/or dry as further explained in this section. In addition, the possibility of having mud or pieces of stool on a surface was mimicked in the laboratory with the use of mashed potato that was inoculated with a known concentration of E. coli (ATCC 25922).

A 0.1 mL of cultured E. coli solution of certain dilution containing approximately 10^9 CFU/mL E. coli was inoculated on the studied surface area. The dilutions were determined to aim for reliable colonies count of 50-200 colonies per plate. To study the recovery rate under wet condition, the surface was immediately swabbed; while to study the recovery under dry condition, the surface was air-dried before it was swabbed. The swab was then returned to its tube which contained 2 mL salt water. The swab tube was shaken thoroughly; the swab was then taken out pressing the tip to the side of the tube wall to squeeze out all trapped salt water remained in the swab tips. Then 0.1 mL aliquots from the swab that was in the bottom of the tube were analysed for the presence of E. coli.

The recovery then was evaluated by comparing the initial amount of E. coli to the swabbed E. coli concentration recovered from the different surface materials and conditions.

6.2.2.2 Phase 2: Field research in the Philippines

The toilet was transported to Abucay Bunkhouse in Tacloban City, the Philippines, a transitional shelter for displaced people affected by the Jolanda Typhoon in December 2013. It is a home for 199 families who lost their houses during the typhoon.

The placement of experimental eSOS Smart Toilet at the transitional shelter was coordinated with an international non-government organisation called Samaritan Purse, who was coordinating all water-sanitation-hygiene (WASH) related activities in the shelter. The community of Abucay Bunkhouse was informed about the research plan and involved in the site selection. The community chose the pedestal-type of user interface and use water for anal cleansing, instead of toilet paper. Because water is scarce, toilet paper was also provided as alternative cleaning method. Training on proper use of the toilet was given to community members who volunteered to participate in the research by using the toilet.

The experimental eSOS Smart Toilet was researched on several sort of studies, social acceptance, characterisation of faecal sludge and urine, as well as testing of all sensors and features. The evaluation of UV-C light effectiveness under real toilet usage was conducted in this phase.

Effectiveness of UV-C disinfection. Based on field considerations combined with the previous UV-C irradiation experiments conducted using both the collimated beam and the UV lamp located at the experimental eSOS Smart Toilet at the facilities of IHE Delft, the exposure time to the UV lamp at the toilet was fixed to 3 minutes. The UV-C light was programmed to switch on as soon as a toilet user leaves the toilet.

The UV-C tests performed at the IHE Delft laboratory were carried out using cultured *E. coli* (ATCC 25922), while the UV-C tests performed at the field research were carried out using naturally occurring coliform bacteria observed at the sampled surfaces i.e. *E. coli* and total coliform. Bacteria, in particular *E. coli* and total coliform, are used as indicator micro-organisms to quantify the effectiveness of UV-C irradiation in a toilet environment. The occurrence of *E. coli* at a surface indicates faecal contamination, while total coliform is a commonly used bacterial indicator of sanitary quality. Although these two bacteria are not the most UV resistant, nor not always pathogenic, they are the most commonly used bacterial indicator across different disinfection methods (e.g. UV, chemicals, etc.) and media (e.g. water, wastewater, air and surface)(Sobsey 1989).Microbial surface samples were taken at different locations in the experimental eSOS Smart Toilet using the standardised swabbing method (as described on Section Standardised Hygiene Swab Test). Samples at five spots i.e. door-handle, water button, shower head, toilet seat and floor, were taken daily, which means that one swab sample is the representation of 24-hours toilet usage on that particular surface spot under the same condition (UV-on or UV-off). The effectiveness of UV-C disinfection was evaluated by comparing the occurrences of coliform bacteria at the toilet evaluated locations/spots when exposing them to the UV-C light; that is, when switching on or off the UV lamp at the toilet.

The disinfection performance of the UV-C lamp was carried out under "real" toilet usage conditions at the emergency camp. Therefore, in addition to set the UV lamp to set on or off, depending on the experimental research design needs, the toilet was exposed to a daily manual cleaning procedure using different cleaning agents to secure a proper (minimum) hygiene toilet conditions for the users. Thus, the UV-C disinfection performance observations on this experiment were also related to the different manual cleaning regimes performed on the toilet during the evaluated research period. The relation between the UV disinfection and the manual cleaning regimes are presented in Table 6-1.

Table 6-1. UV disinfection performance and cleaning regimes

Cleaning Agent	UV status	Observations
Bleach	OFF	5
	ON	8
Detergent	OFF	4
	ON	2
Water	OFF	2
	ON	2

Effectiveness of toilet cleaning. As the cleaning method influences the presence of bacteria on the toilet surfaces, different cleaning method alternatives were evaluated. During the field testing at Abucay bunkhouse, commonly used and available cleaning solutions were evaluated, such as bleach liquid, powder detergent, and water only. The bleach cleaning solution was prepared by diluting commercial bleach (Zonrox brand) which contains 5.25% by weight sodium hypochlorite (NaOCl) as the active ingredient. A commercial powder detergent branded Surf (advertised to be 'anti-bacterial') was used as the detergent cleaning solution. The powder detergent contains cleaning-agents (such as anionic and non-ionic surfactants, and enzymes), water softeners (such as sodium carbonate, and sodium aluminosilicate), fabric whitener, sodium perborate, an anti-redeposition agent, perfume, washer protection agent (sodium silicate), and processing aids (sodium sulfate). This detergent was produced phosphorous free.

The cleaning chemicals were prepared following the manufacturer's recommended dosages for each particular cleaning agent (5 mL of bleach per 1 L water; and 4 grams of detergent powder per 1 L water). Every time a cleaning agent of choice is mixed with 10 L of water to obtain the cleaning solution. The cleaning solution was used for wiping the walls, shower head, door knob, and toilet seat with the aid of a cleaning sponge. Then, the floor was mopped using the cleaning solution. The cleaning was conducted daily following the same order, which was wiping walls, shower head, door knob and finally toilet seat with a sponge (same sponge for all parts), and then a mop for the floor only. Swab samples were taken daily before and after performing the cleaning. The results were then compared to calculate the performance of these cleaning methods on the bacterial removal.

6.3 Results and Discussions

6.3.1 Phase 1: Preliminary research before field testing at IHE Delft campus

6.3.2 UV-C light intensity measurement

The intensity of the UV-C light, evaluated at different locations in the toilet, was measured to determine the light distribution; thus, the UV-C dose at the evaluated locations can be determined. The intensity can be directly related to the UV-C dose. Eighteen different locations were chosen to quantify the light intensity in the toilet. The locations are shown in Figure 6-2. The results are shown in Table 6-2. Note that the reported intensity (irradiance level) values were measured at perpendicular angles of the UV-C light. Some locations at similar distances to the UV-C lamp, such as location number 4 and 5, as well as location number 16 and 17, exhibited slightly different readings regardless. This is possibly considering slightly changes of the sensor angle positioning when performing the measurement. The sensor was handheld; and therefore, the angle with respect to the UV-C lamp was estimated.

These results showed that the UV-C light was distributed throughout the toilet surfaces. The further the spot from the UV-C lamp, the lower the UV-C light intensity. These results are relevance to determine the UV-C dose at different spots inside the experimental eSOS Smart Toilet. The UV-C dose depends both on the UV-C light intensity and on the exposure time. Therefore, it is not possible to get equal UV-C dose at all sampled surface locations.

6.3.3 UV-C light irradiation test at collimated beam and eSOS toilet

The results obtained when exposing laboratory-cultured E. coli to UV-C radiation performed both on the eSOS toilet UV-C lamp at different locations, and on the collimated beam were compared and they are presented in Figure 6-3. The log removal positively correlated with the exposure times and the light intensity, thus UV-C dose. The light intensity for collimated beam and the water-button was the same, and indeed the removal rate corresponded. The *E. coli* inactivation slows down at approximately 10-minutes of exposure time at the evaluated light intensity. On average, it was calculated that there was about 1.1 log inactivation per minute.

Table 6-2 UV-C light intensity at experimental eSOS Smart Toilet

No	Location	Distance from the lamp (cm)	Irradiance level, W/m^2
1	Hanging in the air*, right under the UV-lamp	20	5.60
2	Top wall left	49	1.48×10^{-2}
3	Top wall right	49	1.14×10^{-2}
4	Lower wall left	92	7.70×10^{-1}
5	Lower wall right	92	7.10×10^{-1}
6	S.O.S button	107	5.95×10^{-1}
7	Above door handle	135	5.27×10^{-1}
8	Door handle	136	4.30×10^{-1}
9	Water button	155	3.15×10^{-1}
10	Shower head	167	2.95×10^{-1}
11	Toilet seat right (with lid)	175	2.60×10^{-1}
12	Toilet seat left	176	2.70×10^{-1}
13	Door wall bottom	188	1.88×10^{-1}
14	Floor behind pedestal under shadow	225	3.30×10^{-2}
15	Floor behind pedestal under shadow, near water button	227	2.20×10^{-2}
16	Floor middle near left sidewall	229	1.90×10^{-2}
17	Floor middle near right sidewall	229	1.80×10^{-2}
18	Floor below door	235	1.70×10^{-2}

The working UV-C dose evaluated on this study ranged from 0.01 to 0.19 kJ/m^2, achieving an *E. coli* log inactivation range from 0.66 to 6.57. It seems that in this study, higher *E. coli* reductions could be achieved when compared to literature. Schenk et al. (2011), reported a 7.2 *E. coli* log reduction with a 1.2 kJ/m^2 UV-C dosage; while Guan et al.(2012) reported a lower (0.67 to 1.13 log) reduction with an UV-C dosage of 0.45 to 3.15 kJ/m^2.

In the case of Guan et al., the UV irradiation might have been less effective because of the different working surface. Guan et al. tested the cap of mushroom which is less smooth than the agar surfaces at which this research was conducted. UV treatment is more effective on smoother surfaces (Gardner & Shama 1998).

A study by Schenk et.al (2011) used a different method. They incubated the indicator organism at the stationary growth phase (37°C for approximately 24 hours), and transferred it to a petri dish in a liquid form in a buffer water solution. Then, the enumeration was done using flow cytometry (FCM). FCM calculates the total concentration of bacteria in a liquid sample, while the standard plate count method, as used in this study, inoculated the bacteria colonies after incubation. Therefore FCM does not distinguish the counts by different bacteria, unless the organism is already isolated in the samples. Thus, it is highly possible that FCM reads more bacteria compared to standard plate count.

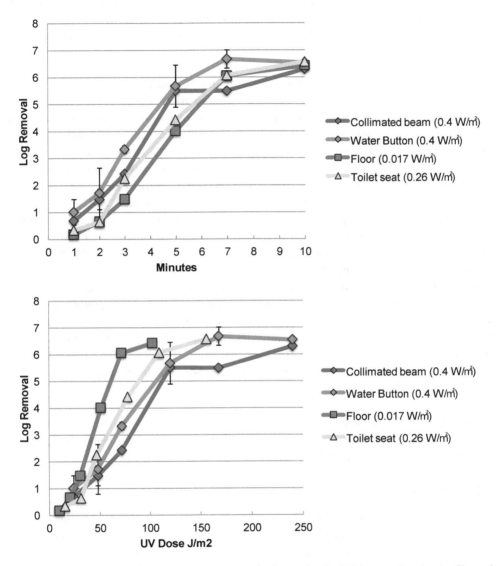

Figure 6-3. *E. coli* log removal at surfaces in the eSOS toilet under the UV lamp and under a collimated beam (with standard deviation as error bar for collimated beam and water button) based on exposure times (top) and UV-C dose (bottom)

Comparing these results with studies on UV irradiation pathogenic micro-organisms, a study using UV dose up to 45 mW.s/cm^2 (equal to 0.45 kJ/m^2) found that the UV dose required to inactivate 99.9% (or 3 log reduction) virus (i.e. enteric viruses poliovirus type 1 and simian rotavirus SA11), pathogenic bacteria (i.e. *Salmonella typhi, Shigella sonnei, Streptococcus faecalis,*

Staphylococcus aureus and _Bacillus subtilis_) and cyst of protozoan _Acanthamoeba castellanii_, was 3 to 4 times, 9 times and 15 times the dose required for inactivation of _E. coli_ respectively (Chang _et al._ 1985). The UV dose required to achieve 3 log removal of _E. coli_ in the study was approximately 70 J/m², which is quite similar to results of this study (see Figure 6-3 right). Pathogenic protozoan _Giardia lamblia_ which has been used as indicator in wastewater treatment, is even less resistant to UV radiation, as more than 4 log reduction can be achieved just with 10 J/m² (Linden _et al._ 2002).

This study found that an exposure time of 10 minutes was enough for achieving a 6.3 logs removal of _E. coli_ at the water button sample locations (see Figure 6-3). The water button location was chosen as referral sampling point in this discussion because of its likeliness to be touched more frequently by unwashed-hands of toilet users. Thus, the water button was likely to be the most critical bacteria transferring point in the toilet. The 10-minutes UV-C exposure would consume energy of about 0.19 kJ/m². In the absence of knowledge on the amount of contamination load and energy consumption of the system (i.e. reliance on solar power supply by the solar cell system) at this point of the research, a 10 minute UV-C exposure was considered to be an ideal exposure time. However, leaving the toilet closed for 10 minutes in the field (that is, avoiding any person to enter the toilet for 10 minutes after each use) was assumed to be too long to guarantee an efficient occupancy by the users added to the concern of excessive power consumption. Considering a shorter exposure time, yet optimum removal capacity, it was found that a 3 minute UV-C exposure at 0.4 W/m², corresponding to approximately 0.057 kJ/m² of UV-C was capable to achieve a 3.33±0.14 log removal of _E. coli_, which already halved the log-removal achieved after 10 minutes (more than triple the 3-minutes time). Therefore 3-minutes exposure time is considered optimal for designing the field test experiments. Hence, a 3-minutes UV-C irradiation was expected to be sufficient to meet both a good disinfection and good toilet occupancy in the field. Three minutes exposure time was then selected as the UV-C setting during the field research.

Performing this evaluation gained the knowledge of the bacteria inactivation capacity of the UV-system installed at the eSOS toilet. It gave insight on how to better designed the experimental conditions at the field evaluation.

6.3.4 Standardisation of swab test

Swab recovery rates performed on different types of surface materials and under different conditions were assessed to quantify the recovery (and indirectly: presence) of bacteria. The results are presented in Table 6-3 as average values with standard deviation. 'Wet surface' refers to the method where a known level of cultured _E. coli_ (diluted with water) was spread on the surface and then immediately swabbed. 'Dry surface' refers to the method where the _E. coli_ solution was allowed to air-dry before the surface was swabbed. The results of the experiment with mashed potato inoculated with cultured _E. coli_ to mimic a lump of faecal matter was not reported here, since there were no sampled surfaces in the field research under such a condition (surfaces with lumps of faecal matters), thus the corresponding recovery rates were not used. For the floor material results, a standard deviation could not be calculated since only one experiment with viable results was obtained.

Table 6-3. Swab recovery rate at different surface materials and conditions

Methods	Recovery rate ([%] as average ± STD)		
	Toilet wall (n=4)	Stainless steel (n=4)	Floor (n=1)
Wet surface	49 ± 3	36 ± 4	25
Dry surface	8 ± 5	5 ± 3	2

The recovery for stainless steel 'water wet' (i.e. 36 ± 4%) is in the same range as reported by Rose et al. (2004) of 41.7 ± 14.6 % using pre-moistened cotton swabs. The wet toilet wall and wet floor results of this study are also in the range of the overall reported swab recovery rates of 25 ± 0.1% (Moore & Griffith 2007). Lower recovery rates were obtained under the dry condition experiments, similar as reported by Lahou and Uyttendaele (2014), which can be due to loss of viability of the micro-organisms by drying. The recovery rates obtained in this experimental phase were used to determine the disinfection/inactivation performance during the field research phase.

Once the swab recovery rates are known, the indicator organism analytical detection limit can be calculated as presented in Table 6-4. These detectable limits are from selected materials and conditions were applied during the field research.

Table 6-4. Detection limit of indicator organism based on different recovery rate

Sampling point	Material and condition	Recovery rate	E. coli detection limit (CFU/100cm²)
Floor	Floor–dry surface	2%	1572
Floor	Floor–wet surface	25%	126
Toilet Seat	Toilet wall–dry surface	8%	384
Shower Head	Stainless steel–dry surface	5%	662
Water Button	Toilet wall–dry surface	8%	384
Door Handle	Stainless steel–dry surface	5%	662

The swab recovery rates assessment was performed to determine the capability of the swab method to quantify indicator bacteria, corresponding to the different surface material in the experimental eSOS Smart Toilet sampled at the field research.

6.3.5 Phase 2: Field research in the Philippines

These experiments were conducted to test the effect of UV-C light on the inactivation of indicator bacteria in the eSOS toilet, taking into account the effect of cleaning types. The interaction of UV-C irradiation and conventional cleaning methods is also discussed. Different conventional cleaning types were also evaluated to conclude the most effective cleaning procedure.

All the samples evaluated in the field research were swabbed surface samples taken in accordance to the pre-standardised method and calculated using the swab recovery rates obtained in the previous laboratory phase evaluation. The wet-surface condition was only found occasionally at the floor during the field research. The rest of the sampling spots were always found to be dry. The swab hygiene test was used in this phase because the researchers were interested in the contamination of the surface by real toilet usage and whether the UV system could work in practice. It is different from the method used during UV tests during laboratory phase which used direct UV exposure of inoculated agar plates, as it aims to assess the UV system capacity as well as comparing it with another UV system at collimated beam.

6.3.6 Effectiveness of UV-C disinfection

The indicator bacteria enumeration data which were obtained through swab sampling during the field-testing has high variability, that the results were best represented by comparing the occurrences rather than reporting the number of bacteria colonies. The occurrences are reported on a daily basis with the UV system on/off as described in Table 6-1. The indicator bacteria occurrences classified into bacteria type and cleaning methods are presented in Figure 6-4.

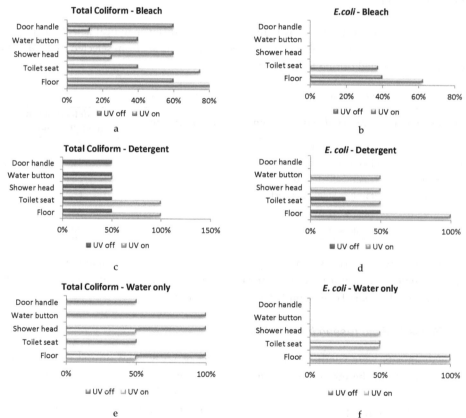

Figure 6-4 Occurrences of total coliforms and *E. coli* when UV was on and off classified considering the different convention cleaning methods simultaneously applied; bleach (a,b), detergent (c,d), water only (e,f)

UV-C light was observed to be effective in the reduction of total coliform at the door handle, water button and shower head, but UV-C was less effective at the toilet seat and floor when cleaned with bleach (Figure 6-4a). While when cleaning with detergent (see Figure 6-4(c)), UV treatment did not show differences of total coliform occurrences and it was not effective at all at toilet seat and floor. While when cleaning with water only, UV treatment was effective at all points (Figure 6-4(e)). Overall, the UV treatment is effective at the door handle, water button and shower head in case of total coliform. UV-C light did not consistently showed any occurrence reduction at toilet seat and floor.

Regarding *E. coli* inactivation, no *E. coli* occurrences were detected at the door handle for all cleaning methods (Figure 6-4(b), 4(d), 4(f)), and occasional occurrences were seen at the water button when cleaned with detergent (Figure 6-4(d)) and at the shower head when cleaned with detergent and water. *E. coli* consistently occurred at the toilet seat and at the floor independent of the cleaning method. Similarly as observed with total coliform, UV-C light did not show to be effective to inactivate *E. coli* at the toilet seat and floor. The almost absence of *E. coli* at the door handle, water button and shower head made the evaluation of UV treatment effectiveness to be inconclusive. However, this result should not discourage the UV efficacy in this study. Since coliforms were reported to be more resistant to UV than *E. coli* and *Salmonella* ((Langlais *et al.* 1991; Yasar *et al.* 2007), sufficiency of UV treatment in removing total coliform indicates the capability to remove *E. coli*.

When observing both indicator bacteria, the UV-C treatment has no impact on bacteria occurrences reduction at the toilet seat and floor locations. Both the floor and the toilet seat are further away from the source of the UV-C light; therefore, these spots received less dose of UV-C irradiation than the door handle, the water button, and the shower head. Furthermore, the insignificant effect of the UV-C light at the floor location could have been caused by the presence of high concentrations of other pollution/bacteria from other sources than human faeces. It was observed that users enter the toilet with contaminated footwear caused by the camp conditions outside (rainwater puddles, dog poop surrounding the toilet, among others). The floor in particular was often found with puddles of muddy water. This surface condition caused UV radiation to be less effective for inactivation of bacteria as they are shielded from UV light by particulate matters (Caron *et al.* 2007; Cantwell & Hofmann 2011). Even the presence of particles in clear water with a turbidity of less than 3 NTU was proven to limit the extent of UV inactivation of indigenous microorganisms (Templeton *et al.* 2005; Cantwell & Hofmann 2008).

A similar explanation may also be valid for the toilet seat, although there was much less visible muddy water found on toilet seat. From the interviews carried out to the toilet users, approximately 10% of interviewees admitted that they were squatting instead of sitting. It implies to the likeliness that while squatting, they contaminated the toilet seat as they put their feet on the pedestal. The high occurrences of *E. coli* and total coliform on the floor and the toilet seat follow the same pattern.

The fact that UV-C inactivation capability is tampered when the microorganisms are protected by particulate matters emphasize the importance of manual cleaning to remove visible dirt and debris that occurred at the floor and toilet seat, as well as other locations in the toilet.

The experiments of UV-C light on and off were conducted for all types of cleaning in order to investigate whether the type of cleaning correlates with the effectiveness of UV-C irradiation on the toilet surface. It was previously observed that *E. coli* only occurred once at water button when UV was on, and twice at shower head, once when UV was on, and once when UV was off (Figure 6-4(b)). These occurrences happened when the toilet was not cleaned using bleach. Effectiveness of different cleaning types is further discussed in the immediate section.

To observe the effect of UV treatment alone without considering the cleaning type, the data was classified into UV-on and UV-off, as presented in Figure 6-5. The observation from this analysis confirms that UV treatment was not effective to reduce both total coliform and *E. coli* at the toilet seat and floor. However, it was shown to be effective for total coliform at the rest of sampling points (i.e. door handle, water button and shower head), and some indication to be effective for *E. coli* at water button. The UV-C effectiveness for *E. coli* at the shower head was questionable attributed to the fact the only one-time *E. coli* occurrence happened when UV was on.

These results demonstrate the capacity of the UV system in the eSOS toilet to disinfect toilet surfaces under real toilet usage. It was found that while UV-C treatment may have been effective for disinfecting surfaces closer to the UV-C light (i.e. door handle, water button and shower head), it is less capable for surfaces that are further away from the UV-C light (i.e. toilet seat and floor). However, the distance was not the only factor determining the efficiency of UV-C. Other factors such as a higher presence of micro-organisms from external source of contamination, together with soil particles that protect the microorganism from the UV-C radiation, also play a role in reducing the UV-C inactivation capability. Therefore multiple measures are suggested. The first recommendation is by increasing the UV-C dose by using more powerful UV-C lamps, or by placing the UV-C lamp closer to the toilet seat and floor. However, increasing the UV-C dose should be done with caution taking into account the UV effects that could accelerate degradation of plastics (Andrady *et al.* 2003; Copinet *et al.* 2004), shortening the shelf life of plastic components of the toilet interior. Another recommendation is applying toilet-management related measures to reduce external contamination e.g. provision of clean sandals only for the use in the toilet, etc., At all cases, a manual cleaning regime remains a necessity to remove visible stains and dirt as UV-C alone is not able to disinfect surfaces in such condition. In the case of contamination of the floor, a better drainage system including addition of floor grating should be considered to allow soil and debris to fall through a drainage channel. Modifications of the design could be applied to frequently touched surfaces to be positioned perpendicularly to the UV-C light, for example in case of the touch surface of the water button. To guarantee cleanliness in all corners and shaded areas is difficult, but by the application of the recommendations above the current system will be improved.

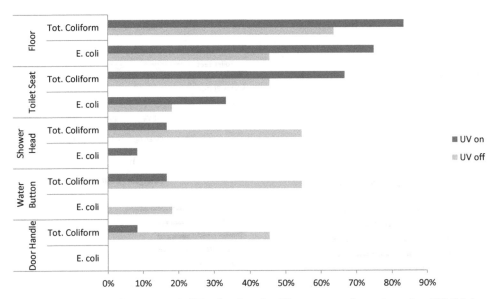

Figure 6-5 Percentage of occurrences of *E. coli* and total coliforms at sampling points when UV-C light on and off

The use of UV-C light as a self-cleaning device in the experimental eSOS Smart Toilet showed some potential in this field research as it has already demonstrated some effectiveness set at a minimum exposure time (i.e. 3-minutes instead of the ideal 10 minutes found on the preliminary phase experiments) corresponding to a minimum power consumption. The current capacity of the system can still be expanded considering that there was abundant power supply as the field research location is situated by the equatorial line receiving strong and consistent solar irradiation to power the toilet. However, reiterating the above-mentioned discussions, increasing UV dosage by adding UV lamps, choose for stronger UV lamp, or positioning the lamp closer to the targeted surfaces – is not the only recommendation to enhance the UV disinfection capacity.

6.3.7 Effectiveness of toilet cleaning regimes

The experimental eSOS Smart Toilet was cleaned daily using three different cleaning solutions. Occurrences of indicator bacteria from the sampled surfaces before and after performing the cleaning activities were compared and are presented in Figure 6-6. The number of observations were as follows $n = 13$, $n = 6$, and $n = 4$, when cleaning with bleach, detergent and water only, respectively.

From Figure 6-6, it was observed that the total coliform occurrences were almost always reduced after cleaning with bleach, except at the shower head where the same occurrence was detected before and after cleaning. The total coliform occurrences were not reduced when cleaning with detergent and water only. At some cases, an increase on the occurrence was observed after cleaning.

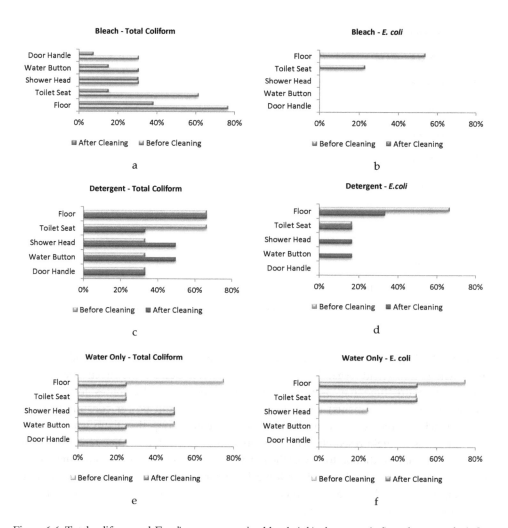

Figure 6-6 Total coliform and *E. coli* occurrence using bleach (a,b), detergent (c,d), and water only (e,f)

E. coli was not observed on the door handle, water button, and shower head sampling points either before or after cleaning when cleaning with bleach. *E. coli* occurrence was not observed after cleaning with bleach at the toilet seat and floor locations. Cleaning with detergent showed some decreases on the *E. coli* occurrence at the floor after cleaning, but it promoted *E. coli* occurrence at the shower button and at the shower head after cleaning. Cleaning with water had some minor effects on the occurrence of *E. coli* at the floor and shower head, but not at toilet seat.

Cleaning with detergent and water-only can not guarantee removal of all bacteria. At some sampling points, even a higher occurrence was observed after cleaning, as shown when cleaning with detergent at the shower head location These results are in agreement with a study by Scott et al.(1984) that observed no reduction of microbial contamination when cleaning using detergent and hot water.

Amongst the three cleaning methods, bleach is the only chemical with disinfecting properties, despite the detergent that was advertised to be 'anti-bacterial'. Detergent is a cleaning agent capable of removing visible dirt, stain and other debris, but it does not really remove pathogen from surfaces (Exner *et al.* 2004). In addition, cleaning with detergent might disseminate pathogens further through the use of contaminated sponge, wipe and or mop (Rutala & Weber 2004). It occurred as the upper surfaces were cleaned using the same sponge, which was repeatedly dunked in the bucket of cleaning solution, possibly re-contaminating the surfaces.

This experiment was conducted to compare the effectiveness of manual cleaning using most commonly used cleaning methods in the field research location. Further, relating to the UV experiments, this experiment was able to demonstrate that self-cleaning feature of eSOS toilet i.e. the UV system should not be the only cleaning mechanism in the eSOS toilet. Effective manual cleaning remain a necessity to guarantee the cleanliness of the toilet interior.

Further research to optimise the cleaning system of the toilet is needed to see the best combination between UV system capacity and manual cleaning frequency. However, this research has provided the baseline of workable UV system and normal manual cleaning.

6.4 Conclusions and recommendations

The followings can be concluded from series of experiments during the preparatory phase and at the field research.

6.4.1 Preparatory phase

- Light intensity characterisation: The UV-C light was distributed all over the toilet interior surfaces where, in general, the further the distance from UV-C lamp, the lower the intensity thus the UV-C dose. The light intensity ranged from 1.14×10^{-2} to 5.6 W/m^2.
- UV-C inactivation: The experiment provided the information necessary to relate the UV system dose to the inactivation capacity of *E. coli*.
- Swab method: The method was required to evaluate the presence of microorganisms on different surface types and at different conditions. The possibility of the swabbing method to enumerate the indicator bacteria is limited by the low recovery rates as observed in this study.

6.4.2 Field Research

- UV disinfection: Total coliform removal was observed on the door handle, water bottom and showerhead for UV doses between 1.88 and 2.74 mW. However, *E. coli* inactivation by UV-C irradiation cannot be confirmed at the evaluated doses, although sufficiency in removing total coliform indicates capability to remove *E. coli* as total coliform is reported to be more resistant to UV than *E. coli*.

- Cleaning regimes: Bleach seems to be the most effective manual cleaning method for both total coliform and *E. coli* inactivation
- The UV system at eSOS Smart Toilet has some potential as it has demonstrated some effectiveness even when set at less than the ideal exposure time.

Based on the findings of this research, the following recommendations for an improved UV system and cleaning strategy in the experimental eSOS Smart Toilet can be made, along with recommendations for further researches.

- To improve the impact of UV disinfection (for example to disinfect the toilet seat and floor), the dose should be increased. This can be done by lengthening the exposure time, and or fixing the UV-C light to closer distance to the toilet seat and floor, or using stronger or more UV-C lamps. In addition, technical modifications that include positioning of the touch surface perpendicularly to the UV-light source and improvement of the floor draining, are recommended to optimise the UV-C disinfection efforts.
- The UV system should not be the only cleaning mechanism in the eSOS toilet, thus effective manual cleaning remains a necessity to guarantee the cleanliness of toilet surface, as well as to enhance UV treatment efficacy.
- Further research to optimise the cleaning system of the eSOS toilet is needed to see the optimum combination between UV system capacity and manual cleaning frequency.

References

Andersen B., Bånrud H., Bøe E., Bjordal O. and Drangsholt F. (2006). Comparison of UV C light and chemicals for disinfection of surfaces in hospital isolation units. *Infection Control* **27**(07), 729-34.

Andrady A. L., Hamid H. S. and Torikai A. (2003). Effects of climate change and UV-B on materials. *Photochemical & Photobiological Sciences* **2**(1), 68-72.

Aycicek H., Oguz U. and Karci K. (2006). Comparison of results of ATP bioluminescence and traditional hygiene swabbing methods for the determination of surface cleanliness at a hospital kitchen. *International Journal of Hygiene and Environmental Health* **209**(2), 203-6.

Barker J. and Jones M. V. (2005). The potential spread of infection caused by aerosol contamination of surfaces after flushing a domestic toilet. *Journal of applied microbiology* **99**(2), 339-47.

Bastable A. and Lamb J. (2012). Innovative designs and approaches in sanitation when responding to challenging and complex humanitarian contexts in urban areas. *Waterlines* **31**(1-2), 67-82.

Bintsis T., Litopoulou-Tzanetaki E. and Robinson R. K. (2000). Existing and potential applications of ultraviolet light in the food industry–a critical review. *Journal of the Science of Food and Agriculture* **80**(6), 637-45.

Bloomfield S. F. and Scott E. (1997). Cross-contamination and infection in the domestic environment and the role of chemical disinfectants. *Journal of applied microbiology* **83**(1), 1-9.

Brown J., Cavill S., Cumming O. and Jeandron A. (2012). Water, sanitation, and hygiene in emergencies: summary review and recommendations for further research. *Waterlines* **31**(1-2), 11-29.

Cantwell R. E. and Hofmann R. (2008). Inactivation of indigenous coliform bacteria in unfiltered surface water by ultraviolet light. *Water research* **42**(10), 2729-35.

Cantwell R. E. and Hofmann R. (2011). Ultraviolet absorption properties of suspended particulate matter in untreated surface waters. *Water research* **45**(3), 1322-8.

Caron E., Chevrefils G., Barbeau B., Payment P. and Prévost M. (2007). Impact of microparticles on UV disinfection of indigenous aerobic spores. *Water research* **41**(19), 4546-56.

Chang J. C., Ossoff S. F., Lobe D. C., Dorfman M. H., Dumais C. M., Qualls R. G. and Johnson J. D. (1985). UV inactivation of pathogenic and indicator microorganisms. *Applied and environmental microbiology* **49**(6), 1361-5.

Connolly M. A., Gayer M., Ryan M. J., Salama P., Spiegel P. and Heymann D. L. (2004). Communicable diseases in complex emergencies: impact and challenges. *The Lancet* **364**(9449), 1974-83.

Copinet A., Bertrand C., Govindin S., Coma V. and Couturier Y. (2004). Effects of ultraviolet light (315 nm), temperature and relative humidity on the degradation of polylactic acid plastic films. *Chemosphere* **55**(5), 763-73.

Cozad A. and Jones R. D. (2003). Disinfection and the prevention of infectious disease. *American Journal of Infection Control* **31**(4), 243-54.

Davidson C. A., Griffith C. J., Peters A. C. and Fielding L. M. (1999). Evaluation of two methods for monitoring surface cleanliness — ATP bioluminescence and traditional hygiene swabbing. *Luminescence* **14**(1), 33-8.

Exner M., Vacata V., Hornei B., Dietlein E. and Gebel J. (2004). Household cleaning and surface disinfection: new insights and strategies. *Journal of Hospital Infection* **56, Supplement 2**, 70-5.

Gardner D. and Shama G. (1998). The kinetics of Bacillus subtilis spore inactivation on filter paper by uv light and uv light in combination with hydrogen peroxide. *Journal of applied microbiology* **84**(4), 633-41.

Gaspar P., Kearns A., Vilar R., Watkins K. and Gomes M. M. M. (2000). A Study of the Effect of Wavelength on Q-Switched Nd:YAG Laser Cleaning of Eighteenth-Century Portuguese Tiles. *Studies in Conservation* **45**(3), 189-200.

Gerba C. P., Wallis C. and Melnick J. L. (1975). Microbiological Hazards of Household Toilets: Droplet Production and the Fate of Residual Organisms. *Applied Microbiology* **30**(2), 229-37.

Griffith C. J., Cooper R. A., Gilmore J., Davies C. and Lewis M. (2000). An evaluation of hospital cleaning regimes and standards. *Journal of Hospital Infection* **45**(1), 19-28.

Guan W., Fan X. and Yan R. (2012). Effects of UV-C treatment on inactivation of Escherichia coli O157: H7, microbial loads, and quality of button mushrooms. *Postharvest Biology and Technology* **64**(1), 119-25.

Johannessen A., Patinet J., Carter W. and Lamb J. (2012). Sustainable sanitation for emergencies and reconstruction situations. In: *Factsheet of Working Group 8*, Sustainable Sanitation Alliance (SuSanA).

Kagan L., Aiello A. and Larson E. (2002). The Role of the Home Environment in the Transmission of Infectious Diseases. *Journal of Community Health* **27**(4), 247-67.

Kouadio I. K., Aljunid S., Kamigaki T., Hammad K. and Oshitani H. (2011). Infectious diseases following natural disasters: prevention and control measures. *Expert Review of Anti-infective Therapy* **10**(1), 95-104.

Kowalski W. (2009). UV surface disinfection. In: *Ultraviolet germicidal irradiation handbook*, Springer, pp. 233-54.

Lahou E. and Uyttendaele M. (2014). Evaluation of three swabbing devices for detection of Listeria monocytogenes on different types of food contact surfaces. *International journal of environmental research and public health* **11**(1), 804-14.

Langlais B., Reckhow D. A., Brink D. R., Foundation A. R. and Compagnie générale des e. (1991). *Ozone in water treatment : application and engineering : cooperative research report.* Lewis Publishers, Chelsea, Mich.

Linden K. G., Shin G.-A., Faubert G., Cairns W. and Sobsey M. D. (2002). UV Disinfection of Giardia lamblia Cysts in Water. *Environmental science & technology* **36**(11), 2519-22.

Moore G. and Griffith C. (2007). Problems associated with traditional hygiene swabbing: the need for in-house standardization. *Journal of applied microbiology* **103**(4), 1090-103.

Nilsen S. K., Dahl I., Jørgensen O. and Schneider T. (2002). Micro-fibre and ultra-micro-fibre cloths, their physical characteristics, cleaning effect, abrasion on surfaces, friction, and wear resistance. *Building and Environment* **37**(12), 1373-8.

Rose L., Jensen B., Peterson A., Banerjee S. N. and Arduino M. J. (2004). Swab materials and Bacillus anthracis spore recovery from nonporous surfaces. *Emerg. Infect. Dis* **10**(6), 1023-9.

Rusin P., Maxwell S. and Gerba C. (2002). Comparative surface-to-hand and fingertip-to-mouth transfer efficiency of gram-positive bacteria, gram-negative bacteria, and phage. *Journal of applied microbiology* **93**(4), 585-92.

Rutala W. A., Gergen M. F. and Weber D. J. (2010). Room decontamination with UV radiation. *Infection Control* **31**(10), 1025-9.

Rutala W. A. and Weber D. J. (2004). The benefits of surface disinfection. *American Journal of Infection Control* **32**(4), 226-31.

Schenk M., Raffellini S., Guerrero S., Blanco G. A. and Alzamora S. M. (2011). Inactivation of Escherichia coli, Listeria innocua and Saccharomyces cerevisiae by UV-C light: study of cell injury by flow cytometry. *LWT-Food science and technology* **44**(1), 191-8.

Scott E., Bloomfield S. F. and Barlow C. (1984). Evaluation of disinfectants in the domestic environment under 'in use' conditions. *Journal of hygiene* **92**(02), 193-203.

Sobsey M. D. (1989). Inactivation of Health-Related Microorganisms in Water by Disinfection Processes. *Water Science and Technology* **21**(3), 179-95.

Stibich M., Stachowiak J., Tanner B., Berkheiser M., Moore L., Raad I. and Chemaly R. F. (2011). Evaluation of a pulsed-xenon ultraviolet room disinfection device for impact on hospital operations and microbial reduction. *Infection Control* **32**(03), 286-8.

Templeton M. R., Andrews R. C. and Hofmann R. (2005). Inactivation of particle-associated viral surrogates by ultraviolet light. *Water research* **39**(15), 3487-500.

Waring S. C. and Brown B. J. (2005). The Threat of Communicable Diseases Following Natural Disasters: A Public Health Response. *Disaster Manag Response* **3**(2), 41-7.

Yasar A., Ahmad N., Latif H. and Amanat Ali Khan A. (2007). Pathogen re-growth in UASB effluent disinfected by UV, O3, H2O2, and advanced oxidation processes. *Ozone: Science and Engineering* **29**(6), 485-92.

Zakaria F., Garcia H., Hooijmans C. and Brdjanovic D. (2015). Decision support system for the provision of emergency sanitation. *Science of The Total Environment* **512**, 645-58.

7

User acceptance of the experimental eSOS™ Smart Toilet

This chapter is published as:

Zakaria F., Thye Y. P., Garcia H., Hooijmans C. Spiegel A. D., and Brdjanovic D. User acceptance of the eSOS™ smart toilet in a temporary settlement in the Philippines. *Water Technology and Practice 12 (4), 832 – 847*

Abstract

An eSOS (emergency Sanitation Operation System) Smart Toilet experimental prototype, aimed at improving the provision of safe sanitation in emergency settings, was field tested in a temporary settlement in Tacloban City, Philippines. The design, usage, and user acceptance of the toilet were all evaluated. Quantitative and qualitative data were collected through interviews and questionnaires, supported by the research-team's observations. The survey results indicated that 98% of users (both first-time users and those who tried it a few times) intended to use the toilet again. There were more features that the users liked than disliked. The in-built water supply and user-operated smart toilet features were liked, but the bad smell was disliked. User-operated smart features were an important factor in user acceptance although they were not the main incentives. Key recommendations are to improve the toilet's design to address the odor and cleanliness issues, make handwashing more convenient, and lower the height of the toilet bowl.

Keywords: smart toilet; field-test; emergencies

7.1 Introduction

During a disaster, a community may be displaced and have limited access to basic necessities such as water, food, shelter and sanitation. Such communities are vulnerable to health-related risks. Disaster responses aim to reduce health impacts and meet the basic needs of the affected communities. The Sphere Handbook (Sphere 2011) provides guidance on the minimum standards required in humanitarian response for those affected to survive and recover, in stable conditions and with dignity. This encompasses the safe disposal of human excreta in order to keep the environment free of human feces and provide adequate toilet facilities. Safe excreta disposal depends on an understanding of needs, including the preferences and cultural habits of the intended users.

It is important to emphasize strongly the importance of user acceptance when providing toilets in emergency situations. Sanitation technologies provided in either emergency or developing country contexts must take user needs into account. This is particularly important for toilets, the user interface of the sanitation chain (Zakaria et al. 2015), which by definition reflects the relationship between toilets and users. User acceptance addresses human preferences and cultural habits. Bichard et al. (2006), reporting on public toilet designs, concluded that both user perceptions and social conventions were key factors for acceptance or rejection. Gyi et al (2013) emphasized that understanding local sanitation behavior was paramount for successful toilet design.

Challenges arise during emergency situations because of the limited time and resources available to consult users or identify their preferences. In such conditions large numbers of toilets must often be deployed quickly, sometimes thousands in a few weeks. In an emergency, solutions that are unfamiliar to users may have to be implemented. However, it is still advisable, where possible, to secure prior approval from users before providing the solution. In practice, humanitarian agencies concurrently promote the use of implemented toilets by complementing implementation with user guides and rigorous hygiene education (Patel et al. 2011; Sphere 2011; Bastable & Lamb 2012), but often without addressing user acceptance.

To minimize the risk of rejection by users during implementation, toilets designed for emergency situations can be field tested in similar settings, to assess their acceptability. Feedback would allow developers to validate or improve designs. This chapter describes an evaluation of user acceptance of an eSOS Smart Toilet – a new design differing from conventional toilets. It was field tested in a temporary settlement in Tacloban City, Philippines.

7.2 Data and methods

A field study was conducted at Abucay Bunkhouse, a temporary settlement in Tacloban City, Philippines, over nine weeks from March to May 2015. At that point, the bunkhouse had been occupied for more than a year, and the residents were using sanitation facilities constructed prior to its opening. Quantitative and qualitative data concerning the test toilet were collected from multiple sources as presented in Table 7-1.

Table 7-1 Sources of data

Activities	Obtained data	Relevance to study objectives
Users registration	Name, body weights, age, household location, and key number	Users identification
Toilet visitors log-sheet	Key number and time in the toilet	Users identification
Users interviews	Users opinion on eSOS toilet (e.g. likes and dislikes, experiences in the toilet, suggestions, rating on smell and cleanliness), how they use the toilet's features, motivation to use	Acceptance, use and preference
Camp-wide questionnaires	Pre-existing sanitation facilities and sanitation practices, overall opinion about eSOS toilet	Context, residents' sentiment towards eSOS toilet, residents' preference, comparison of users and non-users
Toilet's sensors recording	Numbers of daily usage	Context
Research team observation	Smell, state of cleanliness, usages	Triangulation

Before starting the field test, every household in Abucay Bunkhouse was asked whether members would be willing to use the test toilet. Households whose members agreed to use it were issued with numbered access keys. In total, 93 access keys were distributed to 91 households, two households receiving additional access keys on request. Non-resident visitors who chose to use the test toilet were lent an access key by the researchers.

The name, gender, age, and body weight of all those over seven years old who said they wanted to use the toilet were registered. Body weight was used to identify individual users through the automatic data recording system. Users were also asked to record the time and their access key number in a log sheet inside the toilet cubicle whenever they used it. The log sheet and body weight data were used to identify candidates for interviews.

A free t-shirt was offered as an incentive to users who completed at least five log sheet entries. The potential impact of this on the study findings was taken into account during data collection and analysis.

Two surveys were conducted to gather residents' opinions, i.e. the test toilet user interviews and a camp-wide survey. The user interviews, conducted with 70 respondents, utilized a structured questionnaire written and conducted in English. A research team member conducted the interviews, accompanied by a community volunteer to translate questions and answers when interviewees had difficulty. Interview questions enquired about use and perceived conditions of the test toilet, awareness of how the toilet was maintained and by whom, opinions on its features, and suggestions regarding functionality and design.

The camp-wide survey targeted residents of Abucay Bunkhouse regardless of whether they were users or non-users of test toilet. In total, some 126 residents responded to a semi-structured questionnaire. It sought residents' opinions and attitudes towards pre-existing sanitation facilities and to practices before residents moved to the bunkhouse. The questionnaire included a question on opinions concerning the test toilet. The full survey results covering opinions on sanitation facilities before moving to the bunkhouse, etc, are reported by Thye (2016).

7.3 Results

7.3.1 Existing sanitation practices at test site

The Department of Public Works and Highways had constructed two blocks of 63 latrines interspersed with bathrooms (Figure 3-38) when Abucay Bunkhouse opened. Up to nine households (comprising approximately 30 people) shared each operational and accessible latrine. Sets of households were expected to take responsibility for maintaining the latrine their members used.

The latrines were pour-flush units with toilet bowls (Figure 7-1). The pedestal was lower and smaller than European standard pedestals, and wastewater was discharged to a communal septic tank. Before the typhoon disaster in 2015, 52% of the survey respondents had used pedestal toilets and 48% squat toilets.

Figure 7-1 Typical toilet pedestal in use in Abucay Bunkhouse (Photo: F. Zakaria)

Almost all residents reported performing anal cleansing with water after defecation. Only five said that they used toilet paper, which was considered a luxury.

Despite the installation of pour-flush toilets and the practice of cleansing using water, there was no system to distribute water to individual latrines. Instead, people collected water manually from standpipes or a spring, or paid for a private supply directly from the spring to their latrines.

7.3.2 Respondent profile

7.3.2.1 Test toilet users interviews

The 70 user-interview respondents comprised 41 females and 29 males. Of these, 41 had used the toilet regularly and 29 had used it fewer than four times. The respondents also represented a range of age cohorts (Figure 7-2). 'n' in Figure 7-2 refers to the number of respondent per age cohort. The number shown in the bar refers to the percentage proportion of the total respondents. For example: the first bar shows there were 19 respondents in the "7-14" age cohort, where females constituted 17% of the total respondents, and male respondents were 11%.

7.3.2.2 Camp-wide questionnaires

Of the 126 respondents to the camp-wide survey, 39 were already test toilet users and 87 were not.

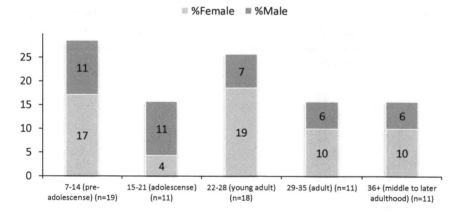

Figure 7-2 Respondent profile

7.3.3 Acceptance

Acceptance or non-acceptance was assessed initially by recording users' expressed inclination to keep using the test toilet once they had used it once or up to three times. All but one of 29 first/few-time test toilet users (97%) said they would keep using it. The one exception explained she preferred her dedicated, shared latrine, simply because of its clear sight line to her house, so she could keep an eye on her infant.

Sentiment towards the test toilet was also evaluated by asking what people liked and disliked particularly about using it. Some 62 respondents gave 111 positive answers. Excluding the 7 respondents who said that they were 'happy with everything', the 104 positive answers were divided into 16 categories (Table 7-2). The seven who said they were 'happy with everything' about the test toilet were excluded, as they did not refer to happiness with any particular toilet aspects. 69 respondents gave 76 answers to 'dislikes' questions. Being less varied than the positive responses, and since 26 (out of the 69; 38%) respondents answered 'no problems', and, so identified nothing they disliked about the test toilet, the remaining 50 genuinely 'negative' answers were divided into just 7 categories.

Most like/dislike categories relate directly to the toilet's design. Some respondents mentioned things concerning specific features while others mentioned only the feature(s) but failed to explain what it was that made them like them. For example, 'likes' Category 7 relates to the toilet bowl achieving water savings because the drop hole does not require flushing. Some categories may also overlap – e.g., Category 4: 'Hi-tech'/automated features could include both Category 2: 'water button- shower head' and Category 6: 'smart key system'.

From Table 7-2 it can be seen that the in-built water supply and the water-button and shower head were aspects most appreciated by users. Next come ease of use, the toilet being 'hi-tech' and cleanliness. The list of 'dislikes' shows that almost half the respondents answered 'smell' as a problem followed by cleanliness issues.

From the camp-wide survey, the majority of both users and non-users, 77% and 64% respectively, were positive about test toilet, commenting frequently on ease of use, attractive appearance and hi-tech features.

Small proportions of user (5%) and non-user (7%) respondents gave negative feedback, while 18% of users and 29% of non-users gave neutral answers such as 'no-comments', and/or mixed answers (positive and negative). Some non-users simply said that they did not use it.

Acceptance of the test toilet by users is related to the features that respondents found particularly attractive or innovative, as can be seen from the five positive issues with the most responses – Figure 7-3. Responses relating to those issues were therefore further evaluated with respect to gender and age. The analysis by age is presented in Figure 7-3. 'n' refers to number of respondents either as a total or particular age cohort. The percentage relates to the number of answers by particular cohort (age cohorts or total). For example, there are 17 respondents age between 7 and 14 years old that responded to 'likes' question. Eight of these 17 – i.e., 47% – said that they liked the 'water-button'.

The in-built water supply and hi-tech features were generally the two most popular features across the age cohorts, both gaining in appreciation with increasing cohort age. Pre-adolescents (7-14 years old) liked the water button/shower far more than any other merits but were not interested in the 'clean' criterion. In contrast, no adolescents (15-21 years old) chose to comment positively on the water button/shower. Ease of use was selected comparably across all respondent groups, excluding pre-adolescents. Turning to gender differentials, men and women appreciated the in-built water supply and hi-tech features almost equally.

Table 7-2 Users' likes and dislikes on the test toilet divided into design and non-design related aspects

Category	Likes	Numbers of answers	Category	Dislikes	Numbers of answers
Design related					
1	Water supply/no need to fetch water	17	1	Odour	23
2	Water button - showerhead	16	2	Dirty/wet	10
3	Easy to use/simplicity	15	3	Related to discomfort: hot during the day, small space, stools visible in the tank, beep sound	9
4	'Hi-tech'/automated features	13	4	Toilet bowl too high	3
5	Clean	12	5	Difficulties with lock	2
6	Smart key/lock system	5	6	No lights at night	2
7	Toilet bowl/no-need to flush/save water	4			
8	The toilet is beautiful, or 'nice'	4			
9	Privacy/security/safety	3			
10	Uniqueness	3			
11	Comfortable (for general interpretation of the word and for aspects not mentioned in other categories)	3			
12	Smells nice	2			
13	Suitable for emergency	1			
Not design related					
14	Accessibility/within distance reach	3	7	Far distance	1
15	Toilet paper	2			
16	The clock (placed to help users fill in the correct time in the log sheet)	1			
	Total answers (respondents)	104 (62)			50 (69)

It was expected that non-acceptance of the toilet would be related to drawbacks perceived by users or the study-site community in general. A list of designer- and researcher- anticipated problems was presented to respondents to determine what they perceived as important. Most frequently listed were the unpleasant odor, a dirty/soiled floor, and water-ponding on the floor (Figure 7-4). Problems associated with odor and cleanliness were assessed in more detail subsequently. There was little difference in perception between women and men concerning the main problems, which was also true across the age cohorts, although pre-adolescents reported the 'water button doesn't work' problem relatively frequently. Interestingly, the toilet bowl being too high was identified across all age cohorts, rather than only by pre-adolescents as expected (because they are shorter than older respondents). This suggests that many adult respondents found the novelty of a pedestal, as against a squatting slab, somewhat disconcerting.

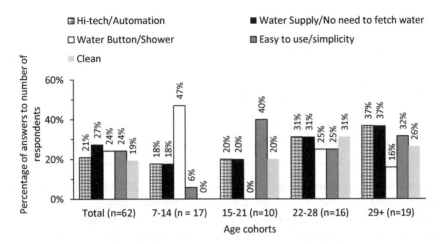

Figure 7-3 Liked features of test toilet, by age cohorts

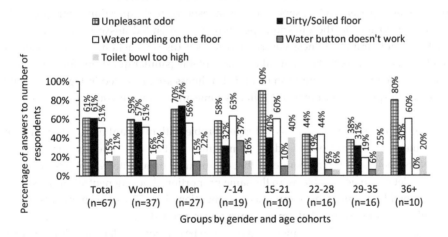

Figure 7-4 Drawbacks experienced in the use of test toilet

In an attempt to confirm factors that might hinder user acceptance, in undertaking both the users' interviews and the camp-wide survey interviews, respondents were asked for suggestions for improvement. The suggestions received and numbers of respondents are listed in Table 7-3.

Table 7-3 Design improvement suggestions received

Category	Suggestions	Users' interviews (n = 69)	Camp wide questionnaire (n = 36)
1	No comment/ok/compliments	36	27
2	Odor-related: air-freshener, ventilation	15	-
3	Cleanliness-related, e.g. more cleaning works, to provide toilet-dedicated slippers, to provide self-help cleaning tools (brush, mop)	13	1
4	Accessories (not odor or cleanliness-related), e.g. bigger space, cooler, soap, manual lighting	7	3
5	Miscellaneous, e.g. instruction translations	1	-
6	More toilets / separate toilets for women and men	-	5

Both surveys yielded similar opinions, apart from category 6 Table 7-3, where only the camp-wide survey generated responses suggesting a preference for more toilets and for separate toilets for women and men. Most respondents to the camp-wide survey had no suggestions and were happy with the test toilet as it stood. However, the user interviews found a number of suggestions were related to improvements related to smells and cleanliness.

In the user interviews, unpleasant odour (smell) and cleanliness were consistently identified as the top problems experienced by users. This was in response to the question where probable problems were listed and in the questions permitting open-ended answers. To clarify the extent to which smell and cleanliness contributed to user dis-satisfaction, user respondents were asked to classify as acceptable, neutral or unacceptable (similar questions for both smell and cleanliness). The results relating to smell and cleanliness are presented in Figure 7-5a and 6-b, respectively. Respondent numbers (n) differ as some answers were void.

It appears that odour was regarded as a bigger problem than cleanliness, since 64% of respondents mentioned unpleasant odour compared to the 33% mentioning lack of cleanliness. Nevertheless, of the respondents that commented on an unpleasant odour, about 36% rated the toilet odour acceptable, with the same proportion choosing neutral, and just 27% rating it unacceptable. Unpleasant odour was also mentioned by 33% of respondents under 'dislikes', with 22% suggesting that improvement was needed. It appears from this that between 22 and 33% of respondents considered odour to be a major problem and thus a source of dissatisfaction among test toilet users.

A similar evaluation showed that test toilet cleanliness was considered less problematic than the bad smell. Some 12% of respondents considered it unacceptable. 49% rated the toilet's cleanliness as acceptable, 38% were neutral.

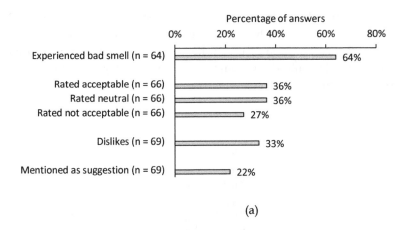

(a)

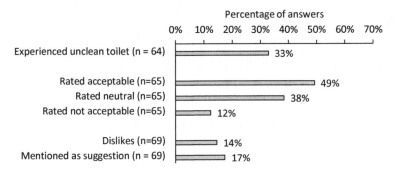

(b)

Figure 7-5 Perception and evaluation on (a) smell and (b) cleanliness at test toilet

7.3.4 Usage

Use of the test toilet was studied to find out whether it complied with the expectations of the designers, and whether the current design and operational principles are such as to promote its use. The in-built sensors recorded daily usage – see Figure 7-6. Usage was monitored for 7 weeks, when there were 662 visits – i.e., 13.5 visits per day on average.

To find out whether users were using the toilet as intended, they were asked about their toilet practices – see Figure 7-7. Questions included whether they sat to urinate, knew how to operate the water button to get water for anal cleansing, aimed into the right part of the bowl when urinating and defecating (small hole for urine, big hole for faeces), closed the toilet lid after use (to reduce odour), washed their hands at the sink behind the cubicle, etc. Most respondents said they knew how to operate the water button (95% female respondents; 88% male respondents) and were aware of the different functions of the holes in the urine-diversion toilet bowl (91% of both gender). The majority (91% of men and 72% of women) also claimed that they closed the lid after use, while noting that they did so only when they remembered.

However, most did not use the hand-washing station (only 36% men and 34% women reported having washed their hands using the sink behind the test toilet). Further questioning revealed that they preferred to wash their hands in their own homes. 97% of women users had no problem with sitting to urinate but only 41% of men were willing to change from their usual behaviour of standing while urinating.

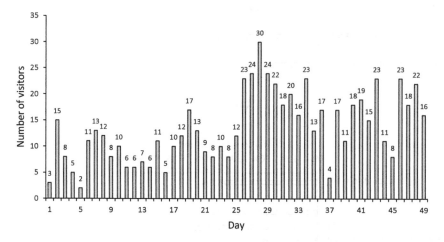

Figure 7-6 Number of daily visits to test toilet during its first 49 days of operation

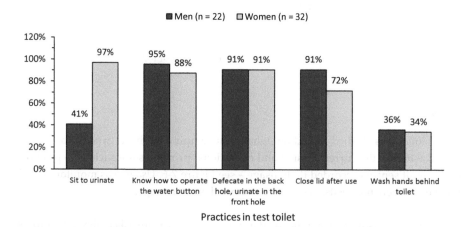

Figure 7-7 Practices in test toilet by gender

Figure 7-8 shows respondents' toilet activity by gender and age. Most pre-adolescents (7-14 years old) reported visiting the test toilet to defecate and urinate, rather than just to urinate, while adolescents (15-21 years old) used it predominantly only to urinate. Those in the higher age cohorts used it for urination or defecation with small preference differences between the two activities.

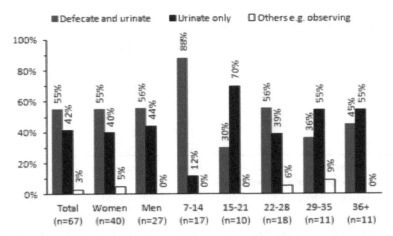

Figure 7-8 User activities in test toilet by gender and age

In order to assess how secure people felt about using the toilet, respondents were asked if they used it alone or in company with another person, and also whether they helped a child who was unable to use it by him-/her-self. Most (nearly 80%) felt confident enough to use the toilet alone, but 11 of 32 women (about 34%) liked company inside the toilet or someone to wait outside. The majority of them (55%) were 7 to 14 year old girls. The few adults who reported needing company explained that it was because they had not used the test toilet previously and wanted somebody to assist them should they encounter difficulties (e.g. not knowing how to get water by operating the water button, or not being able to open the door from inside, etc.).

One respondent reported experiencing a black out while using the test toilet at night (see Table 7-2, 'Dislikes' section). While this did not recur, and the particular user continued using the toilet, it seemed that it constrained use by some other camp residents (Thye 2016), or contributed to the need by some users to be accompanied to the toilet.

7.3.5 Preferences

To assess toilet-users' opinions when comparing the test toilet with other toilets that they normally used, two comparison questions were asked, i.e., whether respondents would choose to use the test toilet rather than other toilets and whether the test toilet was cleaner than other toilets – see Figure 7-9.

The results for both questions showed that almost all users preferred the test toilet (only 2% indicated disagreement). While 22% chose to be neutral, the majority (77% - the sum of 'agree' and 'strongly agree') positively preferred the test toilet.

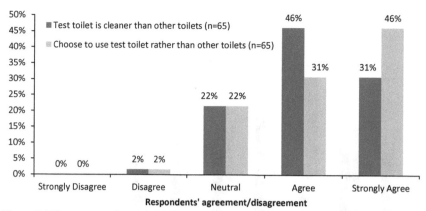

Figure 7-9 Users comparison of eSOS toilet cleanliness and preferences to other toilets

Users' preference for the test toilet was also the main motivation (78% of answers) for choosing to use it (Figure 7-10). Other motivations were 'own toilet in use' (44%), 'no water at own toilet' (44%), and 'wanted the t-shirt' (48%). Preference for the test toilet was the highest motivation in all respondent groups except the adolescents (age 15-21) whose main motivation was to get the 'I-love-eSOS-toilet' t-shirt. It is noted, however, from the log data, that users continued using the test toilet after receiving a t-shirt (i.e. after using it 5-times).

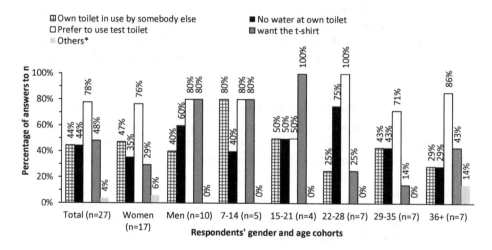

Figure 7-10 Users motivation to use eSOS toilet

Evaluating resident's preferences, Thye (2016)) reported that approximately 58% were satisfied with their own sanitation facility at the same camp. However, it was not clear from the results which factors contributed either to satisfaction or dissatisfaction. Thus interpretation of the results to determine residents' preferences had to be based on the suggestions provided in Table 7-3, without comparison of residents' preferences for their own sanitation facilities.

7.4 Discussion

The study's findings are limited by the way in which the field study was implemented, the short study period, and the small number of respondents. Only one test toilet was installed in a camp where sanitation facilities were already available for all. The mindset and motivation of users in a setting where they could choose to use another toilet can be very different from that in a setting where users can only use one type of toilet. However, the implications of this facet of the study design on the findings have not been taken into account.

Apart from the issues cited above, the overall nine-week study period – of which just seven weeks were used for monitoring the test toilet usage – are insufficient to monitor long-term changes in user acceptance and use, particularly because the novelty of test toilet and the researchers' presence might have increased interest in the short- term.

Thirdly, while 90% of test toilet users responded to the survey, it was difficult to disaggregate findings by gender and age-cohort due to the small sample size.

7.4.1 Acceptance

Some concerns about using the test toilet were observed at first, including worries that users would lock themselves in, that they would break something in the toilet while using it, or that there might be CCTV (closed-circuit television camera) surveillance inside the toilet cubicle. Once these were cleared and they became familiar with the toilet's smart features, most new users became regular users. Judging from users' inclination to continue using the toilet (97% of first/few time(s) users), it was accepted well.

It was feared initially that the unit's smartness could be a factor that might hindered its acceptance, for instance, that residents might feel its use inappropriate, or that they would be concerned that their private business in the toilet might be under surveillance, etc. These fears proved to be ungrounded. In general, the smart features of the toilet were greatly appreciated and users mentioned its hi-tech nature as an attractive feature.

Some smart features were perceived as better than others. The 'water button' received most appreciation and mention. The age cohort analysis shows that pre-adolescents – girls and boys – appreciated the water-button and shower feature most. This was possibly because of their enjoyment of being able to press the button and see water coming out of the shower head. As Harter (1999) and Brinthaupt & Lipka (2002) have indicated, pre-adolescents are typically curious, and tend to be playful and excited by new and uncommon things as they go through a period of dramatic developmental transitions.

Young adults of both genders and older women most appreciated the comfort, and possibly the related benefit that meant that they did not have to fetch water to use the toilet. The drudgery of having to fetch water for their households would have been a common concern for most such respondents.

7.4.2 Relationship between acceptance, use and preferences

For users, the most attractive features of the test toilet were its convenience and the in-built water supply. For surveyed non-users, however, the lack of ownership and sharing with others (whom they might not know) were the main considerations against using the test toilet. While a few people mentioned the idea of having more test toilets to enable separate male and female facilities, this was not widely expressed.

The toilet's water system, including related features like the water button and shower head, was generally used correctly. Use of the urine-diversion toilet interface, however, did not always comply with user's guide instructions. Only 41% of male users sat to urinate as indicated by the toilet's user guide. This supports the results from another study reporting that only 42% of the male respondents sat to urinate in their surveyed urine diversion 'NoMix' Toilet (Lienert *et al.* 2006).

There was no evidence that people did not like using the toilet alone, suggesting that the toilet provided sufficient privacy, protection from surveillance, and lighting. Users also seemed to feel happy using the toilet for defecation and/or urination, indicating that they felt sufficiently comfortable about it.

Community preferences regarding their expectations of a sanitation facility were inconclusive, according to the camp-wide survey. It was concluded, however, that the community was generally satisfied with the available sanitation whilst staying at Abucay Bunkhouse. Analyzing answers from both the user interviews and the camp-wide questionnaire showed that the test toilet's most liked feature was the in-built water supply. This was in a context where the existing camp sanitation facilities lacked a water supply for anal cleansing and flushing. The test toilet, with its water supply provision and non-flushing system, provided comfort for users who did not need to fetch water before going to the toilet.

In summary, the results showed that the Abucay Bunkhouse community, particularly the test toilet users, was positive about the test toilet installed there. Its smart features were appreciated, although they were not the main reason for people liking using it more than the others. The in-built water supply was the major driver and more appealing than, e.g., uniqueness, attractive appearance, etc.

7.4.3 Design improvements

Several suggestions for design improvements have been gathered.

Odour

It appeared that the interior odour problem came mostly from urine, via a faulty urine odour trap, even though the opening was much smaller than that to the faeces tank, and to a smaller extent from urine remaining on the toilet surface or from spills on the floor. The faeces tank did not emit much smell despite its large, uncovered opening. Technical remedies are required for a working urine odour trap. Better ventilation could also be considered.

Cleanliness

Lack of cleanliness was observed exclusively for the toilet floor, which was described by users as 'dirty and wet'. One improvement should focus on better drainage on the floor, perhaps with gridded flooring.

Hand-washing facility

Placement of the hand-wash facility needs reconsideration. The current position, behind the toilet and enclosed by the toilet's fencing, seemed to cause under-utilization. The fencing that had been placed around the toilet to protect the hand-washing station was such that it meant that users needed to go around the toilet and open the fencing gate to get to the hand-washing station. For many, as they explained, it seemed more convenient to go straight home after toilet use and wash there. However, if the fencing were to be removed, the hand-washing facility would have to be placed so that it was well protected from potential damage. At this site, as the researchers observed, children played with the sink and taps, leading to water wastage and potentially to broken facilities.

Toilet bowl

The toilet bowl was higher than deemed necessary by users. The test toilet prototype in this study incorporated a commercially available, portable, urine-diversion toilet, with a higher than normal pedestal. Some people reported this as a problem. To increase user comfort, the pedestal height could be modified to a lower (standard) level.

There were few issues with use of the urine-diversion model. Although there was some hesitation by male users to sit to urinate, this did not jeopardize the pedestal functions. Some respondents even expressed appreciation that this model of toilet reduces water use.

Other improvements

A small number of users mentioned other design-related concerns, including the heat within the cubicle during the day and visible stools in the faeces tank. Better ventilation, as suggested to reduce the odour, may help reduce the temperature, although the outside temperature at the study site could reach 45°C. The PVC faeces tank, which was originally white, was changed on site by covering it with black duct tape to make the faeces less visible. Better remedial action is, however, needed. It may require redesign of the holding tank to preclude users seeing the tank's contents.

7.5 Conclusions

Acceptance

The eSOS Smart Toilet was positively received by users. They said that its main merit was its in-built water supply. Its 'hi-tech' nature was appreciated, with the 'water-button' feature most liked among other user-operated smart features. In general the smart features of the toilet contributed as a factor to acceptance rather than non-acceptance.

The unpleasant odour and lack of cleanliness of test toilet could limit its use. Odour was identified as a bigger problem than cleanliness, with 33% of user respondents mentioning it as a problem compared to 17% concerned with cleanliness.

Use

In general people used the toilet as anticipated by the designers, despite the unconventional features requiring different operation from common toilets. For instance, it was designed with the expectation that users would sit to urinate (it is a urine-diversion toilet with a pedestal). Although a significant number of male users did not do this, the toilet nonetheless remained functional.

Preferences

Preference for the test toilet when compared to other shared toilets was clearly stated as the current users' motivation (78% of all answers) to use it, although the other shared toilets were evaluated as acceptable and an improvement compared to previous sanitation facilities before people lived at this site.

Designs improvements

Concerns expressed during the field test and related to bad odour, lack of cleanliness, difficult of access to the handwashing facility, and the toilet bowl being unusually high all indicate a need for design improvements.

References

Bastable A. and Lamb J. (2012). Innovative designs and approaches in sanitation when responding to challenging and complex humanitarian contexts in urban areas. *Waterlines* 31(1-2), 67-82.

Bichard J., Hanson J. and Greed C. (2006). Away from home (public) toilet design: identifying user wants, needs and aspirations. In: *Designing Accessible Technology*, Springer, pp. 227-36.

Brinthaupt T. M. and Lipka R. P. (2002). *Understanding early adolescent self and identity : applications and interventions*. State University of New York Press, Albany, United States.

Gyi D., Sims R. E. and Gosling E.-Y. (2013). Re-Inventing the Toilet: Capturing user needs. In: *The International conference on Ergonomics & Human Factors 2013,* Anderson M (ed.), CRC Press, Cambridge, UK, pp. 327-8.

Harter S. (1999). *The construction of the self : a developmental perspective.* Guilford Press, New York, United States.

Lienert J., Thiemann K., Kaufmann-Hayoz R. and Larsen T. (2006). Young users accept NoMix toilets-a questionnaire survey on urine source separating toilets in a college in Switzerland. *Water Science & Technology* **54**(11-12), 403-12.

Patel D., Brooks N. and Bastable A. (2011). Excreta disposal in emergencies: Bag and Peepoo trials with internally displaced people in Port-au-Prince. *Waterlines* **30**(1), 61-77.

Sphere P. (2011). *The Sphere Project : humanitarian charter and minimum standards in humanitarian response.* The Sphere Project, Rugby.

Thye Y. P. (2016). *A framework to improve the product development process to achieve more effective sanitation response during emergencies.* Doctoral dissertation, Environmental Engineering, Institut Teknologi Bandung, Bandung, Indonesia.

Zakaria F., Garcia H., Hooijmans C. and Brdjanovic D. (2015). Decision support system for the provision of emergency sanitation. *Science of The Total Environment* **512**, 645-58.

Decision support system for the provision of emergency sanitation

This chapter is published as:

Zakaria F, Garcia H, Hooijmans C, Brdjanovic D. (2015) Decision support system for the provision of emergency sanitation. *Science of the Total Environment* 512: 645-658 (*IF 4.9*)

Abstract

Proper provision of sanitation in emergencies is considered a lifesaving intervention. Without access to sanitation, refugees at emergency camps are at a high risk of contracting diseases. Even the most knowledgeable relief agencies have experienced difficulties providing sanitation alternatives in such challenging scenarios. A computer-based decision support system (DSS) was developed to plan a sanitation response in emergencies. The sanitation alternatives suggested by the DSS are based on a sanitation chain concept that considers different steps in the faecal sludge management, from the toilet or latrine to the safe disposal of faecal matters. The DSS first screens individual sanitation technologies using the user's given input. Remaining sanitation options are then built into a feasible sanitation chain. Subsequently, each technology in the chain is evaluated on a scoring system. Different sanitation chains can later be ranked based on the total evaluation scores. The DSS addresses several deficiencies encountered in the provision of sanitation in emergencies including: the application of standard practices and intuition, the omission of site specific conditions, the limited knowledge exhibited by emergency planners, and the provision of sanitation focused exclusively on the collection step (i.e., just the provision of toilets).

Keywords: Decision support system; sanitation technologies; sanitation chain; emergency sanitation; disasters

8.1 Introduction

This chapter describes a computer-based decision support system (DSS) developed for selecting the most suitable sanitation alternative for emergency situations. The sanitation alternatives suggested by the DSS were defined considering a sanitation chain approach (that is, each sanitation alternative includes excreta disposal, collection, conveyance, treatment, and final disposal or reuse). The computer-based DSS will contribute to ensuring a sustainably operated and maintained sanitation response in emergencies.

Natural and anthropological disasters may lead to the displacement of large numbers of people into temporary settlements or camps. The temporary camps are often overcrowded and contain rudimentary shelters, inadequate safe water and sanitation provision, and a high potential exposure of people (camp residents) to disease vectors. The majority of diseases causing mortality and morbidity in displacement camps (*e.g.*, cholera, diarrhoea, worms, skin irritation, and eye-irritation, among others) have a strong correlation with the state of the sanitation provision at the camps. Without a proper sanitation provision, people living in the displacement camps are at a high risk of contracting diseases.

The word 'sanitation', as well as 'environmental sanitation' could be broadly defined to refer to maintenance hygienic state of certain living environments. This translates into range of activities such as human excreta disposal, household wastewater disposal, vector control as well as solid waste management. However, in the context of emergency where the humanitarian aim is to meet basic sanitation and where the major concern is disease preventions, the word 'sanitation' is considered to have the strongest ties with human excreta disposal and management. Thus for this reason, this study discusses 'sanitation' as excreta disposal management.

The emergency sanitation provision at the emergency camps is predominantly decided by the site planners, which are the corresponding relief agencies together with the local governmental authorities. Due to the many constraints present in an emergency situation, the most commonly selected sanitation alternative has been the simplest possible alternative, limited to onsite decentralized systems, with excavated latrines such as pit latrines or trench latrines being the most popular choice. These basic sanitation alternatives often fail exacerbating even more the problems already encountered in an emergency setting. Failures are due to unstable soils, high water tables, flood-prone areas, locations in which it is not possible to excavate (due to rocky ground conditions, space limitation, and/or land ownership), among others. Such complex emergency scenarios require the provision of sanitation alternatives beyond the old fashioned and problematic latrine, so there is a current need to innovate in the provision of sanitation services considering the complexity commonly observed in emergency situations (Johannessen *et al.* 2012).

After a massive earthquake which caused catastrophic damages in Haiti in 2010, several sanitation alternatives were evaluated, including biodegradable bags (such as *peepoo*-bags), biogas domes, composting latrines, urine-diversion technologies, raised latrines, several pre-fabricated latrines (using diverse materials), and pit lining alternatives to overcome problems of collapsing pits. The majority of these systems were modifications of existing technologies developed for a non-emergency context; however, these alternatives still added some novelty to the field. The performance of these sanitation alternatives was not as satisfactory as expected. As an example, even though the raised latrines were generally well accepted as the best solution for many emergency settlements in Haiti (considering it was impossible to dig pits for latrines), some problems were encountered with this system. The raised latrines were availed and initially maintained by the relief agencies present in Haiti. They required periodical tank emptying to sustain their operation. Once the relief agencies in charge ran out of budget, the latrines were abandoned without a proper closure strategy plan in place (Manilla Arroyo 2014). Several over-spilling latrines were observed. At the time of the cholera outbreak this issue was clearly a major concern. Therefore, the incorporation of sanitation alternatives in an emergency setting without foreseeing the management plan may not result in a sustainable sanitation provision.

The constraints of time and resources placed on the site planners during the planning of the sanitation provision during an emergency situation usually may lead to the use of standard remedies and not optimum solutions (Fenner *et al.* 2007). The lessons learned from previous emergencies situations are not well communicated. That is, the knowledge is kept within the particular relief agencies and/or local authorities. Therefore, site planners often have limited knowledge on the variety of possible sanitation technologies that can be provided leading to unawareness regarding the best possible solution (Mara *et al.* 2007).

8.1.1 Sanitation chain concept

The sanitation provision should be perceived beyond the provision of just toilets or latrines (Verhagen & Ryan 2008; Sparkman 2012), aiming at achieving an overall improvement on public health. The provision of toilets or latrines should be the first step of a series of steps/processes involved in the provision of a complete and solid sanitation alternative. These series of steps are commonly known as the sanitation chain. An interpretation of the sanitation chain concept is described in Figure 8-1. A sanitation chain consists of several individual sanitation processes introduced in a logical order. A sanitation chain includes the following individual processes: (i) processes for excreta disposal and collection from the user-interface (production and capture on Figure 8-1); (ii) processes for excreta conveyance (collection & transport in Figure 8-1); and (iii) processes for treatment until final disposal or reuse (treatment or disposal and reuse in Figure 8-1). Solid waste management is not included in this definition. The sanitation chain as a concept is actually embedded in any sanitation system whether it is off-site (centralised) or on-site (decentralised). However, the actualisation of the concept is more pronounced on the off-site systems rather than on the on-site systems. Off-site systems usually consist of a sequence of several individual sanitation processes where each individual process (representing a step of the sanitation chain) is carried out by a separate sanitation technology. However, on-site systems usually combine several sanitation processes (steps of the sanitation chain) in a single sanitation technology. An example of an on-site system is the simple pit latrine. In a pit latrine (single sanitation technology), the faecal sludge is retained in the pit representing the collection step of the sanitation chain. When no conveyance step is incorporated, the treatment step (*i.e.*, bacterial decomposition) also takes place at the same single sanitation technology (at the collection pit of the pit latrine). The disposal step may also take place at the same single sanitation technology (that is, when the pit is covered once it is totally full). Pit latrines systems perform several steps of the sanitation chain in a single sanitation technology. The incorporation of the sanitation chain concept (or the broader terminology "sanitation value chain") in the literature discussing sanitation systems is fairly recent (*e.g.*, Tilley et al. (2008); Maurer et al. (2012); van Dijk (2012)).

Figure 8-1 Sanitation chain system (modified from Wirmer (2014))

8.1.2 *Review of available sanitation decision support system*

A decision-making support tool is defined as a product that combines information on a user´s given situation with information on available technologies and approaches helping the practitioners to select the best available technology or approach (Palaniappan *et al.* 2008). An accurate DSS will contribute to tackling several of the deficiencies currently observed in the provision of sanitation services in emergencies including: (i) the application of standard practices and intuition from the relief planners in the selection of a particular sanitation option, (ii) the omission of the complex scenarios commonly found in emergency settings, (iii) limited knowledge shown by site planners, and (iv) the misconception that sanitation provision can be achieved by just providing toilets without considering the provision of a whole sanitation chain.

Various support tools have been developed to assist with the selection of the most appropriate sanitation options. SANCHIS (Buuren 2010) recommended the use of a participatory multi-criteria analysis (MCA) to select the most appropriate technology for sustainable drainage and sanitation systems. An exhaustive list of selection criteria for used in MCA for basic sanitation was proposed by Garfi and Ferrer-Marti (2011). Similarly, Katukiza *et al.*, (2010) combined the use of expert opinions with a participatory processes to select most suitable sanitation options in urban slum settings. Tilley *et al.*, (2010) described and classified different sanitation technologies to facilitate an educated decision making sanitation provision process. All of these support systems were developed for non-emergency settings; that is, the complex scenarios commonly introduced by emergency situations were not considered.

Computer programs have been developed in an attempt to facilitate the complex decision making process for the selection of the most appropriate sanitation option. Some of these computer programs include WAWTTAR (Finney & Gerheart 1998), SANEX™ (Loetscher &

Keller 2002), and SETNAWWAT (Sah *et al.* 2010). All of them were designed to be applied in a development (non-emergency) context; therefore, they will not be that accurate when dealing with emergency scenarios.

Some DSS were indeed developed considering the emergency context. They are presented in the form of standard document (*e.g.*, SPHERE standard (Sphere 2004)), technical briefs (*e.g.*, Reed (2010)), technical books (*e.g.*, Harvey *et al.* (2007); Harvey (2002); Wisner and Adam (2002); Frazier (2008)), decision trees (*e.g.*, Fenner *et al.* (2007); Reed (2010); and UNHCR (1999)), and matrices. Most of these decision support tools do not incorporate key aspects when dealing with the provision of sanitation services in emergencies such as universality, inclusion of the latest developments on technologies, and user-friendly interfaces. In addition, none of these DSS are offered in a computer program format. Akvo-WASTE Netherlands (Castellano *et al.* 2011; Akvo.org & WASTE 2012) has recently expanded an online tool that provides users with some specific selection criteria including a preliminary description of sanitation technologies considering the emergency situation context. Even though this tool incorporates computer programming and the emergency context considerations, the tool cannot be considered as a complete DSS at its current stage.

Therefore, there is a need for developing an emergency sanitation DSS in the form of a computer program considering the complex scenarios commonly found in emergency settings. In addition, an accurate DSS should incorporate the provision of a sanitation chain rather than a single sanitation technology. It should serve as an interactive, practical, as well as user-friendly decision-making tool facilitating the selection of the best sanitation option considering the emergency context. These issues were explored in this study leading to a DSS tool described in this paper.

8.2 Methodology

This research was conducted in the following phases: (i) selection of sanitation technologies to be included in the emergency sanitation DSS; (ii) definition of criteria and selection processes; (iii) development of the DSS conceptual framework; and (iv) computer programming.

8.2.1 Selection of sanitation technologies to be included in the DSS

The sanitation technologies included in the selection process are proven technologies that have been used in previous emergencies, or technologies that have a potential to be used in future emergencies. The selected technologies are further classified considering the steps of the sanitation chain concept previously described and presented in Figure 8-1. A specific sanitation chain was defined for this study consisting of the following steps: (i) user interface, (ii) collection, (iii) conveyance, (iv) semi-centralized 1, (v)semi-centralized 2, and (vi) disposal/reuse. The classification of the individual sanitation technologies in the different steps of the sanitation chain are presented in Table 8-1. The technology definitions are mostly based on the Compendium of Sanitation Systems and Technologies by Tilley *et al.* (2014), but certain technologies descriptions that were not included in the compendium are taken from other sources, being mainly reports from relief agencies.

8.2.2 Definition of criteria and selection processes

Various criteria that affect the selection of sanitation technologies were defined. These criteria are technology-specific, site-specific, and socio-culture-specific. Furthermore, the criteria were classified based on how they interfere in the selection process. There are some criteria that eliminate sanitation technologies when the conditions set by the criteria are not met (screening criteria). Other criteria evaluate the suitability of the sanitation technologies to the given scenario (evaluation criteria) by assigning a certain score to each sanitation technology for each defined criterion.

Not all the sanitation technology options at each step of the sanitation chain are compatible. Therefore, additional considerations were incorporated to determine whether a sanitation technology in a particular step of the sanitation chain could work in combination with the sanitation technologies proposed for the rest of the sanitation chain. The compatibility of each sanitation option was then mapped in a matrix specially designed for evaluating compatibility issues of the different sanitation technologies. The compatibility of one sanitation option to another is assessed and given binary value to express whether the pair is compatible or not. The values were based on information obtained from the literature, as well as from the authors' own interpretation.

8.2.3 Development of the DSS conceptual framework

The information flow for the developed DSS is described in Figure 8-2. All the sanitation options go first through an initial screening process. The screening process evaluates all the sanitation options considering the predefined screening criteria and incorporating all the inputs introduced by the users. The screening process results in all the feasible sanitation options for a particular emergency scenario. The users would then be asked to build a sanitation chain combining the feasible options. The sanitation chain would further be evaluated using the subsequent evaluation criteria (incorporating again the users' inputs) resulting in the most suitable sanitation option (in the form of a sanitation chain) for a particular emergency scenario. The evaluation stage is particularly useful for the users to be able to identify the potential advantages and limitations of their chosen sanitation options. The evaluation stage requires the users to score each option to provide basis of quantifying the quality of each option that the users can compare them with other options.

8.2.4 Computer programming

The computer-based DSS was developed using Visual Basic for Applications (VBA) Version 6 in Microsoft Office Excel 2007 (32-bit version). The software is user-friendly, so the DSS can be executed without the need for specific training. The DSS was designed to operate offline, so it can be applied without the need of being connected to the internet. The DSS programme will be downloadable on-line from a specific website.

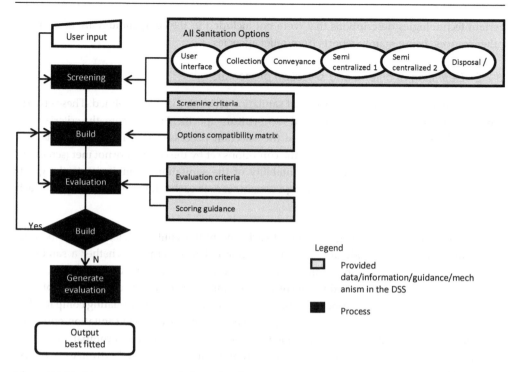

Figure 8-2 Decision support system information flow

8.3 Results and discussion

The results section of this chapter presents the computer based DSS. First the selected suitable sanitation technologies included in the DSS are described. Each selected technology is further classified following the specific components of the sanitation chain. Next, the factors influencing the suitability of all the potential sanitation technologies in the decisionmaking process are identified; they are presented as selection criteria (either screening, or evaluation criteria). Subsequently, the logic followed for conducting the screening and evaluation stages is presented. Finally, the computer based DSS is evaluated comparing the DSS recommendations with the sanitation provision observed in Haiti after the earthquake in 2010.

8.3.1 Selection of sanitation technologies included in the emergency sanitation DSS

The sanitation technologies further classified in the single components of the sanitation chain are presented in Table 8-1.

Table 8-1 Sanitation technologies included in the emergency sanitation DSS (classified in the individual components of the sanitation chain)

Chain components	Technologies	Remarks
(1) User Interface	(101) No User Interface; (102) Drop Hole; (103) Pour Flush; (104) Urine Diversion; and (105) Urinal	Considering that water supply and energy are usually insufficient in addition to unavailability of piped water in emergencies. Cistern and mechanics flushed interfaces are not considered
(2) Collection	(201) Biodegradable Bags; (202) Buckets; (203) Controlled Open Defecation; (204) Shallow Trench Latrines; (205) Deep Trench Latrines; (206) Borehole Latrines; (207) Simple Pit Latrines; (208) Ventilated Improved Pit Latrines; (209) Arborloo; (210) Fossa Alterna; (211) Porta Preta; (212) Septic Tank; (213) Aerobic Filtration (AF); (214) Anaerobic Batch Reactor (ABR); (215) Aqua Privies; (216) Urine Diversion Dehydrated Toilet (UDDT); (217) Urine Diversion Toilet (UDT); (218) Floating Latrines; (219) Raised Latrines; (220) Urine Jerrycan Storage; and (221) Chemical Toilet.	All options are operated without energy requirements
(3) Conveyance	(301) No Emptying/Collection and Transport; (302) Human Powered Emptying/Collection and Transport; (303) Human Powered Emptying/Collection and Motorised Transport; (304) Motorised Emptying and Manual Transport; (305) Motorised Emptying and Transport; and (306) Sewerage	Conveyance includes collection/emptying and transport. Both process technologies can be classified into manual or motorised; thus, the technical options are either manual, motorized, or a combination of the two.
(4) Semi Centralised 1	(401) No Treatment; (402) Co-composting; (403) Planted Drying Beds; (404) Unplanted Drying Beds; (405) Sedimentation/Thickening; (406) Waste Stabilisation Pond (WSP); (407) Surface Flow Constructed Wetlands.	Most of them are primary treatment options that can receive highly concentrated faecal sludge, standalone operated treatment units. They can also act as pre-treatment system for Semi-Centralised 2
(5) Semi Centralised 2	(501) No Treatment; (502) Trickling Filters; (503) Upflow Anaerobic Sludge Blanket (UASB); (504) Membrane Bioreactor (MBR); and (505) Conventional Activated Sludge (CAS)	Treatment options that works in combination with Semi-Centralised 1 systems, or are fed by means of a sewerage system
(6) Disposal and Reuse	(601) Urine Fertilizer; (602) Sludge/Dried Faecal Matter Fertilizer; (603) Burying/Fill and Cover Onsite; (604) Burying/Fill Cover Offsite; and (605) Surface Disposal/Open Dumping	Feasible options with regard to emergencies. Nutrient recovery reuse options are also included

To promote a better understanding of the different sanitation technologies, each option is linked to a pop-up window in the computer program DSS containing brief information of the technology as shown in Figure 8-3. The information includes a general basic description of the technology, advantages, limitations, the applicability of this technology in an emergency situation, and its classification in the sanitation chain. The technology descriptions were taken from the sanitation technology descriptions presented in the compendium by Tilley *et al.* (2014). The information provided in these descriptions may also assist the users to provide a more precise score to each particular technology during the evaluation criteria stage (as described later in this section).

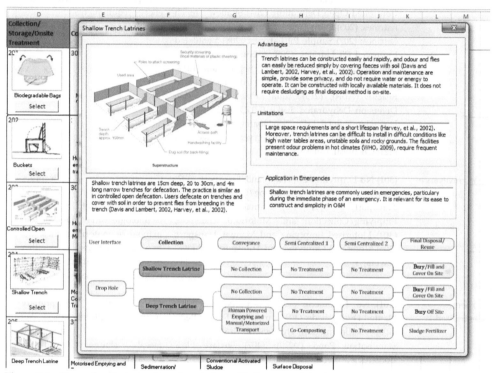

Figure 8-3 - Example of a technology description - Shallow trench latrine

8.3.2 Screening and evaluation

8.3.2.1 Screening and evaluation criteria

The factors (criteria) influencing the suitability of the sanitation technologies in the decision making process were identified and classified as either screening or evaluation criteria as shown in Table 8-2. Site, technology, and/or socio-cultural related aspects were considered when selecting each individual criterion

8.3.2.2 Screening stage

At the screening stage, the 13 screening criteria previously defined and presented in Table 8-2 are incorporated in the form of questions as described in Figure 8-4. Multiple choice options are provided for each question in the form. The user needs to select an answer for each of the questions based on the local situation information or given scenarios at the emergency site.

Table 8-2 Criteria for emergency sanitation DSS classified as either screening or evaluation criteria

Screening criteria	Evaluation criteria
Remaining infrastructure after disaster	Deploy-ability
Water availability to flush	Time to construct or ship
Land availability for latrines cubicle on-site	The use of local material
Possibility to excavate	Technical complexities or requirement of technical
Groundwater table	skills
Eventuality of flooding at the latrine site	Sustainability
Anal cleansing material	Possibility to upgrade
Accessibility by 4W vehicle	Life span (before enquiring new one/de-sludging)
Type of waste stream after collection	Operation and maintenance ease
Energy availability to power de-sludging,	Economical and Environmental Benefit
transport and treatment	Shipping costs
Land availability for off-site treatment	Construction costs
Possibility to excavate at disposal site	Number of people to benefit
Land application/open dumping	Environmental impact
environmentally safe and permitted by local	Potential for end-product re-use
authority	

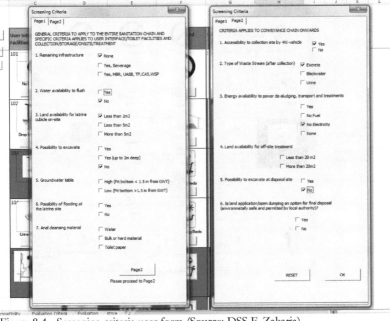

Figure 8-4 - Screening criteria user form (Source: DSS F. Zakaria)

All the choices have a significant impact on the subsequent selected (or discarded) sanitation technologies; that is, each answer discards one or more sanitation technologies which are not suitable under the given conditions or given scenarios. All the sanitation technology options (both the suitable and unsuitable) are presented in the computer program as an Excel spreadsheet as observed on the right hand side of Figure 8-5. As observed in Figure 8-5, the individual steps of the sanitation chain are organized in six sequential columns, and the different individual sanitation technologies are introduced under each individual step of the sanitation chain (column) considering their classification as shown in Table 8-2. The unsuitable options as a result of the screening process are dark red highlighted. The discarded as well as the remaining sanitation technologies after the screening process are presented in Figure 8-5, although for visual clarity reasons, the screen capture only shows a fraction of the entire spreadsheet. As an example, if the option "No" is selected on the Question #2 "Water availability to flush" on the screening criteria form shown in Figure 8-4 (and also at the left hand side on Figure 8-5), this selection discards the "Pour Flush" option at the user interface step of the sanitation chain (first column and third row on Figure 8-5). Subsequently, at the collection step of the sanitation chain (second column on Figure 8-5) sanitation technologies such as "Septic Tank", "Aerobic Filtration (AF)", "Anaerobic Baffled Reactor (ABR)", "Aqua Privy" and "Urine Diversion Toilet" (not shown in Figure 8-5) are all discarded since all of them use pourflush- interfaces. In addition, the "Sewerage" option in the "Conveyance" step of the sanitation chain (third column on Figure 8-5) is also discarded since a sewer does not function without water.

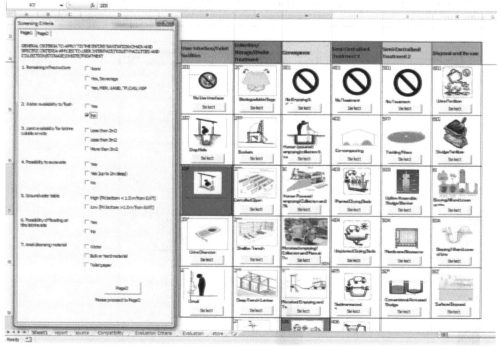

Figure 8-5 - Screen capture after the screening process (unsuitable sanitation options are highlighted in dark red colour) (Source: DSS F. Zakaria)

The complete list of discarded option in relation to one user input when answering each screening criteria can be observed in Table 8-3.

Table 8-3 Discarded sanitation options upon selecting different choices in screening process

Page	Questions	Choices	Discarded Sanitation Options
1	Remaining infrastructure	None	-
		Yes, sewerage	-
		Yes, MBR, UASB, TF, CAS, WSP	-
	Water availability to flush	Yes	-
		No	(103) Pour flush, (212) Septic tank, (213) Aerobic Filtration (AF), (214) Anaerobic Batch Reactor (ABR), (215) Aqua privy, (306) Sewerage
	Land availability for latrine cubicle on-site	Less than 2 m²	(203) Controlled open defecation, (204) Shallow Trench Latrines, (205) Deep Trench Latrines, (212) Septic tank, (213) Aerobic Filtration (AF), (214) Anaerobic Batch Reactor (ABR), (215) Aqua privy
		Less than 5 m²	(203) Controlled open defecation
		More than 5 m²	-
	Possibility to excavate	Yes	-
		Yes (up to 2 m deep)	(205) Deep trench latrines, (206) Borehole latrines
		No	(204) Shallow Trench Latrines; (205) Deep Trench Latrines; (206) Borehole Latrines; (207) Simple Pit Latrines; (208) Ventilated Improved Pit Latrines; (209) Arborloo; (210) Fossa Alterna; (212) Septic tank, (213) Aerobic Filtration (AF), (214) Anaerobic Batch Reactor (ABR), (215) Aqua privy; (217) Urine Diversion Toilet (UDT), (603) Burying/Fill and Cover Onsite
	Groundwater table (GWT)	High (pit bottom < 1.5 m from GWT)	(204) Shallow Trench Latrines; (205) Deep Trench Latrines; (206) Borehole Latrines; (207) Simple Pit Latrines; (208) Ventilated Improved Pit Latrines; (209) Arborloo
		Low (pit bottom > 1.5 m from GWT)	-
	Possibly of flooding at the latrine site	Yes	(203) Controlled open defecation, (204) Shallow Trench Latrines, (205) Deep Trench Latrines, (207) Simple Pit Latrines; (208) Ventilated Improved Pit Latrines; (209) Arborloo
		No	-

Table 8-3 Continued

Page	Questions	Choices	Discarded Sanitation Options
	Anal cleansing material	Water	-
		Bulk or hard material	(103) Pour flush, (212) Septic tank, (213) Aerobic Filtration (AF), (214) Anaerobic Batch Reactor (ABR), (215) Aqua privy, (217) Urine Diversion Toilet (UDT)
		Toilet paper	-
2	Accessibility to collection site by 4W vehicle	Yes	-
		No	(303) Human Powered Emptying/Collection and Motorised Transport; (305) Motorised Emptying and Transport
	Type of waste stream (after collection)	Excreta	(105) Urinal ; (220) Urine Jerrycan Storage, (405) Sedimentation/Thickening; (406) Waste Stabilisation Pond (WSP); (407) Surface Flow Constructed Wetlands;(601) Urine fertilizer
		Blackwater	(105) Urinal ; (220) Urine Jerrycan Storage, (402) Co-composting; (601) Urine fertilizer
		Urine	(402) Co-composting; (403) Planted Drying Beds; (404) Unplanted Drying Beds; (405) Sedimentation/Thickening; (406) Waste Stabilisation Pond (WSP); (407) Surface Flow Constructed Wetlands; (502) Trickling Filters; (503) Upflow Anaerobic Sludge Blanket (UASB); (504) Membrane Bioreactor (MBR); and (505) Conventional Activated Sludge (CAS); (602) Sludge/Dried Faecal Matter Fertilizer
	Energy availability to power de-sludging, transport and treatments	Yes	-
		No fuel	(303) Human Powered Emptying/Collection and Motorised Transport; (304) Motorised Emptying and Manual Transport; (305) Motorised Emptying and Transport
		No electricity	(502) Trickling Filters; (503) Upflow Anaerobic Sludge Blanket (UASB); (504) Membrane Bioreactor (MBR); and (505) Conventional Activated Sludge (CAS)
		None	(303) Human Powered Emptying/Collection and Motorised Transport; (304) Motorised Emptying and Manual Transport; (305) Motorised Emptying and Transport; (502) Trickling Filters; (503) Upflow Anaerobic Sludge Blanket (UASB); (504) Membrane Bioreactor (MBR); and (505) Conventional Activated Sludge (CAS)

Table 8-3 Continued

Page	Questions	Choices	Discarded Sanitation Options
2	Land availability for off-site treatment	Less than 20 m²	(402) Co-composting; (403) Planted Drying Beds; (404) Unplanted Drying Beds; (405) Sedimentation/Thickening; (406) Waste Stabilisation Pond (WSP); (407) Surface Flow Constructed Wetlands; (505) Conventional Activated Sludge (CAS)
	Land availability for off-site treatment	More than 20 m²	-
	Possibility to excavate at disposal site	Yes	-
		No	(604) Burying/Fill Cover Offsite
	Is land application/open dumping an option for final disposal (environmentally safe and permitted by local authority)?	Yes	-
		No	(605) Surface Disposal/Open Dumping

8.3.3 Chain compatibility

After the screening process is finalized, a feasible sanitation chain needs to be selected by choosing one of the remaining (available) sanitation technology options for each step (column) of the sanitation chain. Once a sanitation technology option is chosen (for each step, regardless the order), the computer program runs a compatibility verification test using preloaded information in the form of a compatibility matrix. The computer program automatically discards any incompatible options in response to selection of an option.

The compatibility matrix is built based on the feasible combination of sanitation technologies for a defined sanitation chain. The compatibility is assessed considering how each sanitation option from a specific step of the sanitation chain affects the selection of other sanitation options from other steps of the sanitation chain. As an example, (as shown in Figure 8-6) if the sanitation option "biodegradable bags" is selected in the collection step of the sanitation chain, automatically the computer program determines the compatible sanitation options for the other steps of the sanitation chain and discards all the sanitation options that are not feasible. That is, it would be feasible to have a biodegradable bag option at the collection step of the sanitation chain without the need of having a collection system at the conveyance step of the sanitation chain (first option shown at the very top of Figure 8-6). Consequently, this option leads to an on-site disposal (*i.e.*, bury on-site). Analyzing the second example provided at the bottom of Figure 8-6 (raised latrines), it is observed that when the sanitation option "raised latrines" is selected at the collection step of the sanitation chain, a collection system (or a tank-emptying option) needs to be incorporated in the conveyance step of the sanitation chain to make the entire chain feasible. The feasible sanitation chain pathways for all the listed sanitation technologies were determined in flow charts similar to the one presented in Figure 8-6.

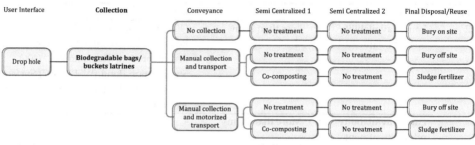

Biodegradable bags and bucket latrines system

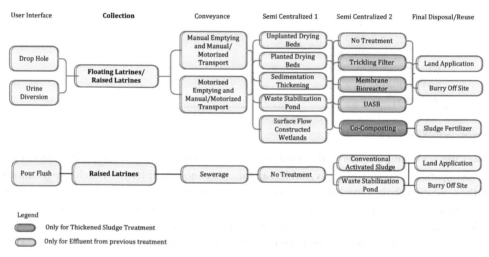

Raised latrines and floating latrines system

Figure 8-6 - Potential chain options when considering "Biodegradable bags/bucket latrines system" (Top) and "Raised latrine and floating latrines system" (Bottom) at the collection step of the sanitation chain (Source: F Zakaria)

All the feasible sanitation chain pathways for each single sanitation technology were translated into a binary coded matrix, used as input to the DSS program. When a certain sanitation technology is selected, the computer program automatically verifies the compatibility using that binary code matrix. Subsequently, the computer program discards all the incompatible options.

Figure 8-7 shows a screen capture of the computer program (Excel spreadsheet) after the screening and the chain compatibility verification processes are completed. The options that were discarded as a consequence of the screening process are dark red highlighted, while incompatible options (after going through the compatibility verification process) are light red highlighted. The highlighted sanitation options can no longer be selected. Figure 8-7 shows a continuation of the same example started when describing the screening stage. For that example, due to the unavailability of water (decided at the screening phase), the "Pour Flush" sanitation option was excluded (dark red highlighted) at the user interface step of the sanitation chain. Subsequently, if the "Biodegradable Bag" sanitation technology is selected at the "Collection" step of the sanitation chain (second column of Figure 8-7), the rest of the sanitation technologies under the same step "Collection" of the sanitation chain are discarded (light red highlighted). In addition, other incompatible sanitation technologies such as the "Urine Diversion" at the step "User Interface" of the sanitation chain, as well as the "Motorised Emptying-" sanitation technology at the "Conveyance" step of the sanitation chain are discarded (light red highlighted).

Figure 8-7 Compatibility verification process (Source: DSS F. Zakaria)

Afterwards, the selection of the sanitation alternatives can be finalized by selecting the remaining sanitation technologies for the rest of the steps (columns) of the sanitation chain. Selected sanitation chains consisting of six different individual sanitation technologies can be created as shown in Figure 8-8 below.

Different chains can be implemented at the same site should they complement and serve different waste stream or different group of users. For example the DSS users might decide to use UD system, thus need to plan for different conveyance-treatment and disposal of urine and excreta. They might also decide to provide a biodegradable-bags-system to serve the elderly and people with special need that reside in parts of one emergency settlement

Figure 8-8 Selected sanitation chain after completing the screening and compatibility verification processes (Source: DSS F. Zakaria)

8.3.4 Evaluation stage

The selected sanitation chain can be then evaluated by applying predefined evaluation criteria. As previously described in Table 8-2, the evaluation criteria are grouped into three categories *i.e.* deploy-ability, sustainability, and economical and environmental benefit. At this evaluation step each single sanitation technology describing the selected sanitation chain needs to be scored. The scoring system ranges from "0" to "5" indicating how well each specific sanitation technology meets the predefined criteria. A complete description with the scoring criteria is presented in Table 8-4. The DSS calculates the final (total) score for each evaluated sanitation chain.

The grouping of evaluation criteria into three categories is attributed to the many contributing factors in evaluating one sanitation technology. These factors often share the same objectives that they can be grouped into one criterion. The grouping of these aspects would enhance the practicality where there would not be too many criteria to be scored individually. Thus it maintains the DSS' aim of being user-friendly, where the users' are given sufficient dose of scoring responsibility.

Table 8-4 Scoring guide for the evaluation criteria

	0	1	2	3	4	5
Deployability	• It takes very long time and process to avail the option on the desired location • the option does not use any local materials • the option requires special equipment and technical skill to avail	• It takes quite long time and process to avail the option on the desired location • the option use almost no local material • the option requires high degree of technical complexities (special equipment and technical complexities)	• It takes some times and process to avail the option on the desired location • the option use little local material • the option requires some degree of technical complexities (special equipment and technical complexities)	• It takes some times and process to avail the option on the desired location • the option use some local material • the option requires some degree of technical complexities (special equipment and technical complexities)	• It takes little times and process to avail the option on the desired location • the option use mainly local material • the option requires little technical complexities (special equipment and technical complexities)	• It takes no times and process to avail the option on the desired location • the option use entirely local material • the option requires no technical complexities (special equipment and technical complexities)
Sustainability	• It is impossible to upgrade the option • the option has very short life span - where it needs continuous replacement and services to be maintained • the option is very complicated to operate and to maintain	• It is remotely possible to upgrade the option • the option has short life span - where it needs continuous replacement and services to be maintained • the option is very complicated to operate and to maintain	• It is possible with some complications to upgrade the option • the option has quite short life span - where it needs continuous replacement and services to be maintained • the option is complicated to operate and to maintain	• It is quite possible with to upgrade the option • the option has considerable lengthy life span - until it needs replacement and services to be maintained • the option is quite easy to operate and to maintain	• It is possible with to upgrade the option • the option has long life span -until it needs replacement and services to be maintained • the option is easy to operate and to maintain	• It is highly possible with to upgrade the option • the option has very long life span - until it needs replacement and services to be maintained • the option is very easy to operate and to maintain
Economical and environmental benefit	• the option is very costly to avail • the option benefits very few people • the option has negative environmental impact • there is no possibility of by product reuse	• the option is costly to avail • the option benefits few people • the option has negative environmental impact • there is limited possibility of by-product reuse	• the option is somehow costly to avail • the option benefits limited number of people • the option has negative environmental impact • there is little possibility of by product reuse	• the option is within considerable cost to avail • the option benefits considerable number of people • the option has negative environmental impact to some extent • there is some possibility of by product reuse	• the option is cheap to avail • the option benefits plenty people • the option has no negative environmental impact • there is good possibility of by-product reuse	• the option is very cheap to avail • the option benefits many people • the option has positive environmental impact • there is high possibility of by-product reuse

Figure 8-9 shows two scoring examples (scoring capture screens on the computer based DSS) applied to the sanitation chain described in Figure 8-8 (top of Figure 8-9), as well as to the sanitation chain example with the raised latrine component described in Figure 8-6 (bottom of Figure 8-9). Some conflicting issues may arise when scoring the sanitation technologies of the chain as follows. As an example, three main aspects need to be simultaneously considered (as described in Table 8-4) when scoring the "Deployability" evaluation criterion including time, the use of local material, and the need for special equipment and technical skills. Some contradictory information for the same sanitation technology may add complexity to the scoring process. For instance, a certain sanitation technology may be deployable in a long period of time (low score), uses mostly local material (high score), and has a low requirement on equipment and technical skills (high score); therefore, priority should be given to the particular aspect that influences the selection the most. That is, for this particular example either the time, or the usage of local material, or the requirement of equipment and technical skill needs to be prioritized. The priority needs to be consistent throughout other options when scoring the same category, also when scoring the next sets of sanitation chains, so that the scores are all comparable.

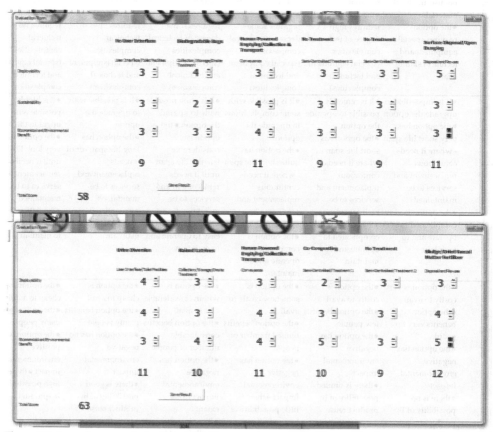

Figure 8-9 A scoring example for the evaluation stage is provided. Top: evaluation stage scoring for the biodegradable bags system example covered in Figure 8; and Bottom: Evaluation stage scoring for the raised latrine system example shown in Figure 8-6. (Source: DSS F. Zakaria)

The results of every single selected and evaluated sanitation chain can be saved, and the complete evaluation process can be restarted from the very beginning selecting a completely different and new sanitation chain. After evaluating all desired sanitation chains, a final report can be generated. The final report compares all the selected sanitation chains (including their final scores) as shown in Figure 8-10. Up to 20 different sanitation chains can be compared in the same report.

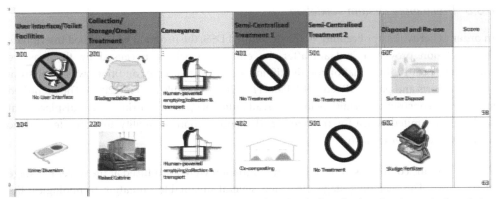

Figure 8-10 Example of a final report following showing the final evaluation for the sanitation chains previously discussed and evaluated in Figure 8-9 (Source: DSS F. Zakaria)

The final report shown in Figure 8-10 compares two feasible sanitation chains as previously discussed. One chain is provided with a biodegradable-bag at the collection step of the sanitation chain, while the other chain is provided with a raised latrines system. The computer-based DSS concluded that first sanitation chain (biodegradable bag) does not require a toilet structure; that is, no user interface is needed. The used bags can be collected using manual collection and transportation, and can be disposed in selected disposal sites without the need for treatment. The second sanitation chain (raised latrines) is provided with a urine-diversion user interface dividing the waste streams into urine and excreta. The excreta (collected in a collection tank) can be regularly emptied using manual emptying and transportation equipment. Then, the collected excreta can be composted at a composting facility, and the resulting product can be used as a fertilizer. The final score obtained for the biodegradable bag and for the raised-latrine chains was of 58 and 63, respectively. That is, the raised-latrine chain can be considered a better sanitation alternative than the biodegradable bag chain. For this particular examples (-referring back to the evaluation scores described in Figure 8-9), the raised latrines chain alternative scored much higher on the 'Sustainability' evaluation criteria (mainly due to the composting and reuse of the excreta as a fertilizer).

The evaluation stage is subject to the DSS user consideration towards certain technologies. The score for one technology may significantly differ when scored by different users. Nevertheless, the provision of both a clear description of the evaluation criteria, as well as a scoring guidance may help users to score the different technologies as accurately as possible. In addition, it is recommended to conduct collective scoring by a team, rather than by individuals to achieve a more objective and proportional scoring.

The final report is the output of the DSS. The final report aims at comparing the advantages and limitations of potential sanitation alternatives by means of quantifying the advantages and limitations with certain scoring system. This comparison may provide users with an elaborated and systematic approach to select the most suitable sanitation alternatives to the given scenarios.

8.3.5 Preliminary system validation

In order to verify the applicability of the developed DSS, the program was applied to evaluate the sanitation provision in Haiti after the emergency situation caused by the massive earthquake in 2010. Information for identifying local or site specific conditions was taken from several reports and papers (Reed 2010; Patel *et al.* 2011; Bastable & Lamb 2012). The identified site-specific constraints in Haiti include unavailability of space for latrines at the displacement settlements, limited space for treatment and disposal, non-possibility to excavate, and no availability of water to flush.

The site-specific conditions at the emergency situation were introduced at the screening stage of the DSS. The screening stage discarded several unsuitable sanitation technologies; the following sanitation technologies were found suitable: (i) no user interface, drop hole, and urine diversion (at the "User Interface" step of the sanitation chain); (ii) biodegradable bags/bucket latrines, porta-preta, floating latrines, raised latrines, and chemical toilets (at the "Collection" step); (iii) all the sanitation technologies options except Sewerage (at the "Conveyance" step); (iv) No Treatment, Co-composting, and planted and unplanted drying beds (at the "Semi Centralized Treatment 1" step); (v) only the No Treatment option (at the "Semi Centralized Treatment 2 step"); and (vi) Sludge Fertilizer and Surface Disposal (at the "Disposal and Reuse" step).

The DSS was able to narrow down all the available sanitation technologies suggesting suitable sanitation options considering the specific emergency scenario. As previously discussed in the Introduction Section, all the screened sanitation technologies were actually in use in the Haiti emergency. Therefore, the DSS yields similar results regarding potential sanitation technologies that can be applied compared to those actually applied by the relief agencies at the emergency site. However, the DSS goes one step further by both suggesting a sanitation chain rather than a single sanitation technology, and by also evaluating and ranking (by scoring all the sanitation technologies in a chain) the feasible resulting sanitation chains.

After the screening stage is finalized several sanitation chains can be proposed. For this particular case, since the relief agencies do not normally want to deal with the maintenance of sanitation facilities, the sanitation technology "No Emptying & Collection" at the "Conveyance" step of the sanitation chain was initially selected. However, the compatibility verification test discarded all the possible sanitation options at the "Disposal and Reuse" step of the sanitation chain, yielding no feasible sanitation chains as observed in Figure 8-11. Therefore, another option was needed to be selected including a different sanitation technology at the

"Conveyance" step of the sanitation chain; that is, it would not be possible to skip some sort of maintenance of the collection facilities by the relief agencies.

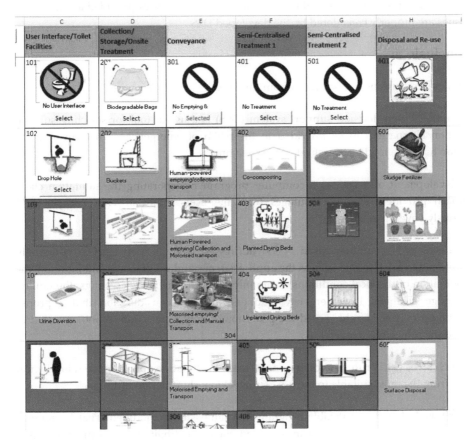

Figure 8-11. Screen capture of the DSS when the "No-Emptying & Collection" option is selected at the "conveyance" step of the sanitation chain (Source: DSS F. Zakaria)

Subsequently, the "Human-powered emptying & collection" sanitation technology was selected at the "Conveyance" step of the sanitation chain. After the program ran the compatibility verification test, the rest of the feasible options for the sanitation chain were selected. The selected sanitation chain was identical to the sanitation chain presented in Figure 8-8 (i.e., the sanitation chain with the biodegradable-bags sanitation technology at the "Collection" step of the sanitation chain). This chain was already evaluated as shown in Figure 8-9, and presented in the final evaluation report in Figure 8-10.

The DSS indeed went one step further than the relief agencies decision at the emergency site, and provided an entire sanitation chain consisting of several compatible sanitation technologies properly evaluated. One of the key design features of the present DSS is the consideration of the entire sanitation chain for the sanitation services to be provided and not just the provision of toilets (collection step of the chain). As was discussed in the introduction of this study, one of the main critical issues with the sanitation technologies provided at the emergency situation in Haiti by the relief agencies was not considering any "Conveyance" and/or "Disposal and Re-use" technologies for the sanitation services provided at the camps. These issues are properly considered by the developed DSS.

8.4 Conclusions

The following conclusions can be drawn from this study:

- A computer-based decision support system (DSS) is a useful tool for selecting suitable sanitation alternatives to be provided in the realm of emergencies when an accurate decision has to be made in the shortest possible time.

- The development of a DSS as a computer program incorporating the sanitation chain approach provides several advantages from the users' point of view compared to previously developed DSSs.

- The developed computer-based DSS addresses several deficiencies commonly encountered in existing DSS such as the application of standard practices and intuition, the consideration of the local conditions commonly found in emergency settings, the limited knowledge exhibited by emergency planners, and the consideration that a proper sanitation provision should exceed the provision of only a collection technology (that is, beyond the provision of toilets).

- The DSS can be easily run providing up to 20 feasible sanitation options ranked in a logical order (from the most suitable to the least suitable) in a short period of time.

- The DSS is designed as a flexible program that can easily be modified. That is, more or different sanitation technologies can be added, the compatibility matrix can be modified to satisfy emergency planner's special needs, and both the screening and evaluation criteria can be changed. In addition, the computer-based DSS is flexible considering that each different user can introduce his or her own inputs depending on personal evaluation of the particular situation.

- The final decision regarding the provision of the most appropriate sanitation alternative entirely depends on the user. The DSS is thought to be a resource to help on the decision-making process.

- Considering the preliminary validation on the Haiti's past emergency situation, it can be concluded that the DSS provides realistic results.

- The DSS is considered a valuable tool for selecting appropriate sanitation services addressing challenging emergency sanitations. Further research is needed to completely validate this tool using data either from past or current emergencies including information related to emergency preparatory activities.

8.5 Acknowledgement

The DSS is developed under the project 'Stimulating local innovation on sanitation for the urban poor in sub-Saharan Africa and South-East Asia' financed by Bill & Melinda Gates Foundation. Special thanks to Mloelya Mwambu from Tanzania that worked on developing this DSS during its early stage as her MSc thesis.

References

Akvo.org and WASTE (2012). The Sanitation Decision Support Tool. http://waste-dev.akvo.org/dst/sanitation/ (accessed 5 September 2014.

Bastable A. and Lamb J. (2012). Innovative designs and approaches in sanitation when responding to challenging and complex humanitarian contexts in urban areas. *Waterlines* **31**(1-2), 67-82.

Buuren J. v. (2010). SANitation CHoice Involving Stakeholders: a participatory multi-criteria method for drainage and sanitation system selection in developing cities applied in Ho Chi Minh City, Vietnam. *Wageningen university*.

Castellano D., de Bruijne G., Maessen S. and Mels A. (2011). Modelling Chaos? Sanitation Options; Support and Communication Tool. *Water Practice and Technology* **6**(3).

Fenner R. A., Guthrie P. M. and Piano E. (2007). Process selection for sanitation systems and wastewater treatment in refugee camps during disaster-relief situations. *Water and Environment Journal* **21**(4), 252-64.

Finney B. and Gerheart R. (1998). A User's Manual for WAWTTAR. *Environmental Resources Engineering, Humboldt State University*, 70.

Frazier C. (2008). Water, sanitation and hygiene in emergencies. In: *Public health guide in emergencies* Rand EC (ed.), International Federation of Red Cross and Red Crescent Societies ; Johns Hopkins School of Hygiene and Public Health, Geneva.

Garfi M. and Ferrer-Marti L. (2011). Decision-making criteria and indicators for water and sanitation projects in developing countries. *Water Science & Technology* **64**(1), 83-101.

Harvey P. (2007). *Excreta disposal in emergencies : a field manual*. Water, Engineering and Development Centre (WEDC), Loughborough University.

Harvey P., Baghri S. and Reed B. (2002). *Emergency sanitation: assessment and programme design*. WEDC, Loughborough University.

Johannessen A., Patinet J., Carter W. and Lamb J. (2012). Sustainable sanitation for emergencies and reconstruction situations. In: *Factsheet of Working Group 8*, Sustainable Sanitation Alliance (SuSanA).

Katukiza A. Y., Ronteltap M., Oleja A., Niwagaba C. B., Kansiime F. and Lens P. N. L. (2010). Selection of sustainable sanitation technologies for urban slums — A case of Bwaise III in Kampala, Uganda. *Science of The Total Environment* **409**(1), 52-62.

Loetscher T. and Keller J. (2002). A decision support system for selecting sanitation systems in developing countries. *Socio-Economic Planning Sciences* **36**(4), 267-90.

Manilla Arroyo D. (2014). Blurred lines: accountability and responsibility in post-earthquake Haiti. *Medicine, Conflict and Survival* **30**(2), 110-32.

Mara D., Drangert J., Anh N. V., Tonderski A., Gulyas H. and Tonderski K. (2007). Selection of sustainable sanitation arrangements. *Water Policy* **9**(3), 305.

Maurer M., Bufardi A., Tilley E., Zurbrügg C. and Truffer B. (2012). A compatibility-based procedure designed to generate potential sanitation system alternatives. *Journal of Environmental Management* **104**(0), 51-61.

Palaniappan M., Lang M., Gleick P. H. and Pacific I. (2008). A review of decision-making support tools in the water, sanitation, and hygiene sector. http://www.pacinst.org/reports/WASH_tool/WASH_decisionmaking_tools.pdf.

Patel D., Brooks N. and Bastable A. (2011). Excreta disposal in emergencies: Bag and Peepoo trials with internally displaced people in Port-au-Prince. *Waterlines* **30**(1), 61-77.

Reed B. (2010). *Emergency Excreta Disposal Standards and Options for Haiti*, DINEPA; Global WASH Cluster.

Sah L., Rosseau D. P. L. and Van Der Steen P. (2010). Selection tool for natural wastewater treatment systems. In: *SWITCH Document*. Software edn, SWITCH.

Sparkman D. (2012). More than just counting toilets: The complexities of monitoring for sustainability in sanitation. *Waterlines* **31**(4), 260-71.

Sphere P. (2004). Humanitarian charter and minimum standards in disaster response.

Tilley E., Supply W. and Council S. C. (2008). *Compendium of sanitation systems and technologies*. Eawag Dübendorf, Switzerland.

Tilley E., Zurbrügg C. and Lüthi C. (2010). A flowstream approach for sustainable sanitation systems. In: *Social Perspectives on the Sanitation Challenge*, Springer, pp. 69-86.

Tilley E. A., Ulrich L., Lüthi C., Reymond P. and Zurbrügg C. (2014). Compendium of sanitation systems and technologies. http://www.eawag.ch/organisation/abteilungen/sandec/publikationen/publications_sesp/downloads_sesp/compendium_high.pdf.

UNHCR (1999). *Handbook for emergencies*. UNHCR, Geneva.

van Dijk M. P. (2012). Sanitation in Developing Countries: Innovative Solutions in a Value Chain Framework. In.

Verhagen J. and Ryan P. (2008). Sanitation Services for the Urban Poor: Symposium Background Paper. In: *IRC Symposium: Sanitation for The Urban Poor - Partnership and Governance*, Delft, The Netherlands.

Wirmer B. (2014). *A functional approach to guide sustainable innovations in the sanitation chain*. Master of Science, Innovation Sciences, Eindhoven University of Technology, Utrecht.

Wisner B. A. J. W. H. O. (2002). Sanitation. In: *Environmental health in emergencies and disasters a practical guide*, World Health Organization, Geneva, pp. 127-47.

Development and validation of a financial flow simulator for the sanitation value chain

Based on:

Claire Furlong, Fiona Zakaria, Damir Brdjanovic, 2018. Development and validation of a financial flow simulator for the sanitation value chain (*In preparation*)

Abstract

The Sustainable Development Goals embrace the concept of faecal sludge management (FSM), moving beyond the provision of only toilets. Hence the financing of all sanitation services from the toilet (containment) to reuse (sanitation value chain) need to be considered if a system is to be considered financially sustainable. FSM requires different financing strategies compared to traditional sewer-based systems, due to a fragmented service chain that due to different service providers and organisational arrangements. A tool has been developed (eSOS Monitor®) to simulate the financial flows along and within the sanitation value chain (SVC). It was developed to enable the users to explore and optimise the financial sustainability across SVC or at a particular part of the SVC. In this chapter, the tool was tested and validated using data from Nonthaburi, Thailand (baseline scenario). As the system in Nonhaburi system is reliant on budget support, several scenarios were modelled using different financial flow models, with the aim to recover the operational and capital expenditure. The results from the models showed that transport and emptying combined, are financially sustainable with an emptying fee of $15, but become unsustainable when a discharge fee is introduced. The treatment process becomes sustainable when a sanitation tax of $50 was introduced, which then becomes the budget support for the treatment procoess. This study successfully demonstrates how the eSOS Monitor can be used to explore different FFMs for a case study area and how it can be used to optimise the financial sustainability across an SVC.

Keywords: Nonthaburi, Thailand, sanitation, business model, eSOS Monitor

9.1 Introduction

A majority of the world's population use onsite sanitation that is 2.7 billion people, and this number is expected to increase to 4.9 billion by 2030 (Cairns-Smith et al., 2014). Onsite sanitation is sanitation systems which store or treat excreta close to the point where it is generated, e.g. pit latrines, septic tanks, etc. In the MDG period (2000-2015) the focus was getting people onto the sanitation ladder, by building latrines and other onsite sanitation systems (OSS). At the start of the MDG period, little consideration was given to the long-term management of OSS, as it was assumed that people would eventually progress to sewer based sanitation. This was an unrealistic goal, and the field of faecal sludge management (FSM) came into the spotlight. Faecal sludge (sludge from OSS) needs to be managed through a series of stages, which replicate what happens in a well maintained and operated networked sanitation system. This is called the sanitation value chain (SVC) (Figure 8-1).

The SDGs have embraced this systems approach (Figure 8-1) as it calls for "safely managed sanitation services", this goes beyond the provision of toilets and embraces FSM. There is a huge diversity of service providers not only within in each part of this system (Figure 8-1), but also along the chain bridging different parts, i.e. NGOs, governments, the informal, private and public sectors. These service providers are linked not only by the physical flow of material along the chain, but also by the transfer of cash between stakeholders and service providers. Therefore the SVC needs to be financially viable if it is to be sustainable.

Several financial decision support tools have been developed for FSM, a majority of them focus from emptying to treatment, due to the capital and operation cost of the user interface and containment being borne by the users. The earlier tools incorporated financial aspects as part of overall sustainability such as SANEX ™ (2000), later financial aspects became a major focus for technology selection (as in WHICHSAN and NEWSAN simulator (Branfield & Still 2009). These tools used capital expenditure (CAPEX) and operational expenditure (OPEX) to assess the financial feasibility of different technology options. Now the focus is on exploring the lifecycle cost across the SVC (Financial Analysis Tool for Urban Sanitation (Campos et al. 2012), WASHCost (Cowling et al. 2013) and FSM Technical and Financial Toolkit (WASHCost 2012). They include financial assumptions such as loan conditions, and in the case of Technical and Financial Toolkit (Ross et al. 2016) a balance sheet for the lifecycle of the project is generated. Due to the enabling environment surrounding sanitation some tools are context specific, such as Financial Analysis Tool for Urban Sanitation (Ross et al. 2016) which has been developed and tested in Bangladesh, and SANIPLAN (Cowling et al. 2013) which has been designed for Indian municipalities. SANIPLAN (2016) (PAS-India 2015) is a comprehensive tool which covers financing and provides an overview of the different financing options across the complete SVC (including the user interface and containment). It enables local governments to review the financial impact of different improvements across the SVC and incorporates a dual licensing and sanitation tax model for the financial flows, which is India specific. It is the only tool which incorporates a financial flow model. Although the complexity of financial decision support tools has grown, to our knowledge, none have incorporated a function that allows the impact of different financial flow models to be assessed across the SVC.

The financial complexity of service provision in FSM is highlighted in a review of 44 business models (a business model is defined as how a company does business, including the value proposition, creation and delivery, delivery and capture carried out by PAS India (PAS-India 2015) i.e. models for emptying and transport of faecal sludge or resource recovery, although they are impacted by the financial and physical flows along the whole SVC (Bocken *et al.* 2014)container-based sanitation (Rao *et al.* 2016)the use of these systems is so far limited.

Financial flows (transfers) are a part of the value capture segment of the business model framework (Rao *et al.* 2016) and are one of the simplest ways to explore financial sustainability. The concept is also known as capital or cash, flow or transfer analysis. This method has been used to assess the financial stability at all levels ranging from companies to sectors e.g. the water, sanitation and hygiene sector (Bocken *et al.* 2014) up to country level (Trémolet *et al.* 2012). It is the analysis of capital (money) movements in and out of an entity, in the case of this study the sections of and across the SVC.

Due to the complex nature of FSM and service delivery, a number financial flow models (FFMs) have already been developed (OECD 2017). These models are divided into two groups, with and without subsidy or budget support (Steiner *et al.* 2003). The financial transactions included in FFMs for FSM can be seen below in Table 9-1, it is noted that different authors use different terms. The five most common FFMs are identified in Figure 9-1.

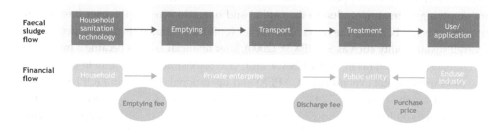

   a.   Model 1: Discrete collection and treatment model

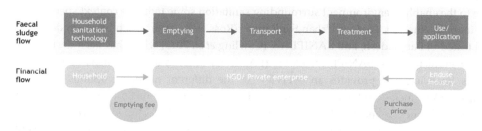

   b.   Model 2: Integrated treatment and collection model

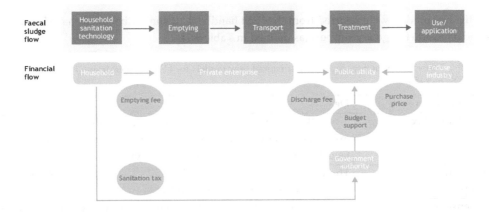

c.   Model 3: Parallel tax and discharge free model

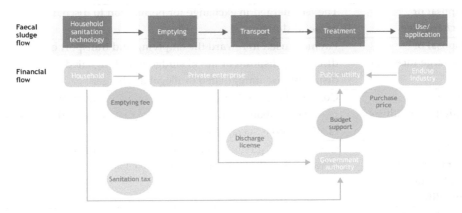

d.   Model 4: Duel licensing and sanitation tax model

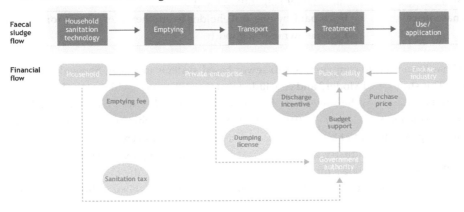

e.   Model 5: Incentive discharge model (Steiner *et al.* 2003)

Figure 9-1 Five most common financial flow model in sanitation (Source: (Strande *et al.* 2014b)

All the five models were referred from the FSM handbook. It should be noted that Models 3, 4 and 5 include budgetary support (as defined in Table 9-1).

Table 9-1 Financial transactions included in financial flow models for FSM adapted from (Tilley & Dodane 2014), (Steiner *et al.* 2003).

Financial transaction	Description
Sanitation tax	Fee collected either once or at regular intervals, which is paid in exchange for environmental services such as connection removal of faecal sludge, or any combination of services.
Collection or Emptying fee	The fee that is charged at the household level for removing faecal sludge from the onsite sanitation technology.
License fee	A financial instrument used to control the number and quality of emptying and transport enterprises that are allowed to discharge faecal sludge at the faecal sludge treatment plant.
Disposal or Discharge fee	The fee charged in exchange for permission to discharge or dispose of faecal sludge.
Disposal or Discharge incentive	Payment used to reward the emptying and transport enterprises for discharge/disposal of faecal sludge in a designated location and to disincentivise unregulated or illegal discharge.
CAPEX or Capital expenditure	Costs that are paid once, at the beginning of the project to cover all materials, labour and associated expenses needed to build the facilities and associated infrastructure.
OPEX or Operation and maintenance expenditure	Costs paid regularly and continually until the service life of the infrastructure/equipment has been reached.
Budget support or Subsidy	Cash transfers between stakeholders to partly or fully cover one stakeholder's operating budget.
Purchase price	Price paid by one stakeholder to another in exchange for becoming the sole owner of a good.

A FFM has been used to compare FSM to sewer-based systems in Dakar, Senegal(Dodane *et al.* 2012). This study used Model 3 and concluded that the OPEX and CAPEX of the FSM system were significantly lower the sewer based system. The financial flows of 44 individual FSM businesses were mapped(Rao *et al.* 2016), but no tool exists to aid this process or to enable the users to compare different FFMs. The modelling of financial flows across the sanitation value chain address the SDG Industry Matrix call to "...apply modelling expertise to help develop financially sustainable models for water projects, using fees and tariff structures which reflect future costs and manage usage while subsidising connections and consumption for the poor" (Rao *et al.* 2016).

This chapter documents the development and validation of an FFM simulator (known as eSOS Monitor), which covers the entire SVC. The FFMs simulated are those found in (UN-Global-

Compact & KPMG-International 2015). The simulator has been developed to enable the user to either explore different FFMs for one SVC scenario or to compare several SVC scenarios using a single FFM. Both approaches aim to explore and optimise the financial sustainability across SVC or at a particular part of the SVC depending on the objective of the user.

9.2 Methodology

9.2.1 eSOS Monitor Development

This simulator uses an add-on of a decision support tool for sanitation technology selection developed by IHE Delft (Zakaria *et al.* 2015). The technology selection tool is used to build the SVC for a particular area. Then for each technology, the CAPEX and OPEX were estimated using previous knowledge and experience, as well potential revenue streams. The CAPEX and OPEX for each component are broken down in detail, e.g. the CAPEX for user interface (defined as the superstructure of the toilet) includes sub-components such as construction materials, transport of materials and construction, whereas the OPEX of this component includes the cost of cleaning materials and water. All of these default values can be changed when real data become available. Additional financial information can also be added for each component of the SVC, such as the budget support available, sanitation tax, cooperative income tax, depreciation etc. Each component of the SVC is linked with the previous and subsequent components, ensuring the flow of information and connectivity between all components. To establish and reasonably describe such links was the most challenging part of the eSOS Monitor development. Table 9-2 provides the summary of input and output data of each SVC component and its linkage with the subsequent step in the process.

The FS flow analysis in an SVC is difficult to attain in reality as the SVC components are fragmented. The eSOS Monitor addresses this gap by monitoring the FS flow from the toilet to treatment or reuse. The eSOS Monitor has been equipped with location map and GPS tracker as well as a separate real-time transportation module. Figure 9-2 shows the screen capture of location of toilets in the study case.

Table 9-2 Summary of input, output and links to next SVC component

SVC component	Input	Output	Data flow to the next SVC component
User interface	<ul><li>Type of user interface</li><li>Number of toilets (individual/shared/communal)</li><li>Number of people per toilet per day</li><li>Amount of faeces, urine, excreta, water usage, black water (BW) generated per day</li><li>CAPEX per unit</li><li>OPEX per unit</li><li>Unit cost of water, electricity, labour</li></ul>	<ul><li>Amount of faeces, urine, excreta, water and BW from the toilet use per day</li><li>Total CAPEX per day, month, year</li><li>Total OPEX per day, month, year</li></ul>	Amount of faeces, urine, excreta, water and BW from the toilet use per day

Table 9-2 Continued

SVC component	Input	Output	Data flow to the next SVC component

Containment	• Type of containment
	• Containment specification:
	  • Holding capacity
	  • Number of containment
	  • Fs accumulation factor
	  • Output streams (sludge/urine/biogas/humus/others)
	  • Sale price of valuable streams (e.g. Biogas, humus)
	• Capex per unit
	• Opex per unit
		• Amount of FS/BW/urine accumulated per containment unit
		• Emptying frequency
		• Number of emptying events per year
		• Total capex per day, month, year
		• Total opex per day, month, year
		• Revenues (from valuable end products)
			Amount of fs/bw/urine collected

Emptying	• Type of emptying • Number of units • Emptying capacity (e.g. M3 FS pumped/hr) • Emptying fee • CAPEX per unit (purchase cost) • OPEX per unit (labour, fuel/energy, technical maintenance, tax, business operation overhead) • Revenue	• Amount of FS emptied per day (need to be consistent with amount of FS accumulated in containment) • Total CAPEX per day, month, year • Total OPEX per day, month, year • Revenues (from emptying service)	Amount of FS emptied
Transport	• Type of transport • Carrying capacity of transportation unit • CAPEX (purchase cost) • OPEX (labour, fuel/energy, technical maintenance, tax, business operation overhead) • Revenue	• Amount of FS transported • Total CAPEX per day, month, year • Total OPEX per day, month, year	Amount of FS transported
Treatment	• Type of treatment • Design capacity • Amount of FS received at the treatment (have to be consistent with amount of FS transported) • CAPEX (construction costs, land requisition) • OPEX (labour, fuel/energy, technical maintenance, chemical, tax, business operation overhead) • Discharge fee (if applicable) • Revenue	• Amount of FS treated • Total CAPEX per day, month, year • Total OPEX per day, month, year • Amount of end products	Amount of end products

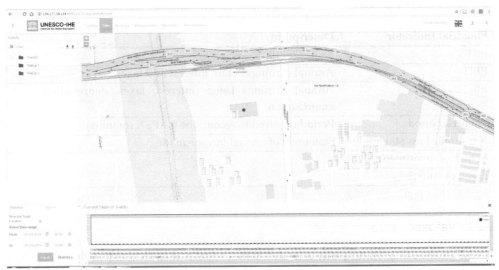

Figure 9-2 Location of toilets in Nonthaburi study case (Source: eSOS Monitor)

Together, GPS tracker and the real-time transportation module, would enable the user to locate sanitation facilities, track the emptying or collection vehicles, calculate transport distances and calculate the most efficient transport route. This would allow fuel consumption and real-time costs for collection vehicles based on their movements to be calculated. To be able to use this transportation module, all vehicles should be equipped with GPS tracker that is linked to the eSOS Monitor. Hence eSOS Monitor can be utilised to monitor real-time operation that includes the dynamic operation of FS transportation (fuel consumptions and time spent according to movements), as well as to evaluate the financial transfer in the SVC. Since the GPS tracker is not yet in application of this study, validation of transportation module is not included. Hence, this chapter focuses on the evaluation of eSOS Monitor as financial flow simulator (FFS) assuming static transport. The eSOS Monitor software was built using programming softwares Java 8 for backend, AngularJS for frontend and PostrgreSQL 9.4 database.

The FFS includes the five most commonly observed FSM financial models (Strande *et al.* 2014a). The CAPEX, OPEX and revenue (Table 9-1) are calculated for each part of the SVC and the chain as a whole, as is the financial indicators listed in Table 9-3. These financial indicators were selected in the FFS to give an estimation for service providers of how much it would cost on a yearly basis to serve a definitive population or number of households. Those indicators are most useful for business users assessing business viability in FSM. Subsequently, it also seeks to give an estimate on how much subsidies or taxation required to keep service providers financially afloat.

The parameters reported in this chapter are CAPEX, OPEX, revenue, EBITDA (as neither tax nor depreciation was used in the simulations), payback period, net profit/loss, breakeven point and the return of investments (ROI), as these were deemed to be the most important parameters for this analysis.

Table 9-3 Financial indicators generated in the eSOS Monitor

Financial Indicator	Description
EBT	Annual earnings before tax
EBIT	Annual earnings before interest and taxes
EBITDA	Annual earnings before interest, taxes, depreciation, and amortisation
Payback period	Period required to recoup the CAPEX (months)
Net profit/Loss statement	Statement of annual income made
Return of Investment (ROI)	
Break-even-point	

9.2.2 Validation area

The Nonthaburi City Municipality is located in the central part of Thailand north of the capital Bangkok. It is also considered to be a suburb of Bangkok due to its closeness, and it is linked to the capital by Bangkok's public transport system. The city covers areas of 38.9m² and has a population of 256,457 people (129,597 households) (Harada *et al.* 2015). The population density in this city is the second highest in Thailand (Harada *et al.* 2015). As in the rest of Thailand, the population is reliant on onsite sanitation predominantly single and double ring cesspools which generally have open bottoms to allow the effluent to infiltrate (Harada *et al.* 2015) or the effluent is discharged to open drains or sewers (Aecom & Eawag 2010; Harada *et al.* 2015). Due to the Public Health Act (1992) the responsibility for septage management lies with the local government (Harada *et al.* 2015). The faecal sludge treatment plant in Nonthaburi gained royal support; this is why there is a vast amount of data on faecal sludge management in this city. Nonthaburi FS collection system has also been considered good as it was estimated to succeed to collect more than 80% of total generated FS (Harada *et al.* 2015).

9.2.3 Validation process

The validation area was chosen based on the availability of detailed data. Nonthaburi City Municipality (NCM) in Thailand chosen for that reason. Hence it has been extensively studied, and detailed data is available (AECOM et al., 2010; AIT, 2012; CSE, 2011; Harada et al., 2015). Additionally, the financial flows of the SVC in NCM have already been modelled (AIT, 2015), so a comparison of results can be drawn. The detailed data that was used to generate the FFMs for Nonthaburi can be seen below in Table 9-4.

This data was used to generate a baseline scenario. In order to evaluate the baseline scenario using different fees and financial instruments, four scenarios were plotted (Table 9-5). These scenarios were evaluated aiming at full cost recovery. The amount of fees and tax proposed was based on the assumption that it would still be affordable for the fee bearer.

Table 9-4 Data from NCM

	User interface & Containment	Emptying & Transport	Treatment	Reuse	Data Source
Stakeholder	Households	Municipality	Municipality	Consumer	AIT,2015
Type of Technology	Combination of pour flush – cistern flush with water for anal cleansing, connected to either single pit or septic tank	Desludging truck	Anaerobic baffled reactor tanks	Compost	Tilley and Dodane (2014), (Aecom & Eawag 2010)
Quantity - Capacity	Volume of pit or tank 1.5 m³	2 x 4m³ trucks 2 x 6m³ trucks	31 units rotated on a daily basis	6,767 ton per year	AIT,2015
CAPEX	Default parameters used $0*	$312,000	$950,000 land acquisition $850,000 construction cost		AIT,2015
OPEX (Annual)	Data not available	$59,000 per year Personnel 52% Fuel 25% Maintenance 23%	$115,000 per year Personnel 11% Materials 19% Maintenance 61% Utilities 9%		AIT,2012; AIT,2015; ACEOM et al., 2010
Current emptying fees	$13 per pit or tank				AIT,2015; Harada et al., 2015
Annual revenues		$67,000		$9,000	AIT,2015
Additional information	0.26 kg of FS generated per person per day Emptying frequency once every 1-2 years				AIT,2015
	Number of households 16,000 Household size 2 Number of toilets 16,000 10 L of water used per person per day 5900 pits or tanks emptied per year Monthly salary of workers $275 (minimum wage)				Calculated

*) assumed included in the construction cost of the house

Table 9-5 Scenarios tested

Scenario	FFM	Parameters changed
1	2	Emptying fee = $20 (per 1.5 m³)
2	1	Emptying fee = $15 (per 1.5 m³) Discharge fee = $5 (per 1.5 m³)
3	3	Emptying fee = $15 (per 1.5 m³) Annual sanitation tax = $50 (per household) Discharge fee = $5 (per 1.5 m³)
4	4	Emptying fee = $15(per 1.5 m³) Annual sanitation tax = $50 (per household) Annual discharge license = $50 (per vehicle)

9.3 Results and discussion

Learning from NCM study case, the faecal sludge and FFM follows Model 2 (Table 9-1), as shown in Figure 9-3, emptying, transport and treatment are managed by public utilities which is part of the municipality. To ensure the clarity of the financial transfer between the components, emptying and transport is taken as one component, while treatment is another component. Since these components are entirely funded by municipality, there is no financial transfer between emptying and transport and treatment in the baseline case.

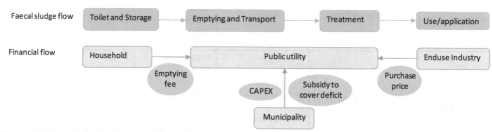

Figure 9-3 Faecal sludge flow and financial flow of Nonthaburi (modified from AIT, 2015)

The AIT case study of financial flows of Nonthaburi only covered the SVC from emptying to reuse and only uses the OPEX and revenue to simulate the financial flows (AIT, 2015). The justification for this is that the municipality covered the CAPEX for emptying, transport and treatment. In their financial flow, it could be seen that there as an annual net loss of –$98,000 which the municipality is required to fund annually (Figure 9-4).

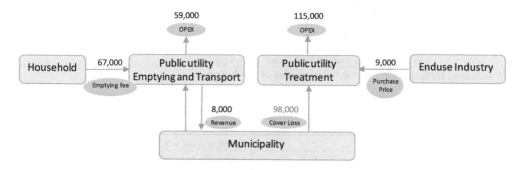

Figure 9-4 Financial flow for Nonthaburi Municipality (Modified from AIT, 2015)

The Emptying and Transport component was financially self-sustaining since it gained sufficient funds from the emptying fee to cover its OPEX. A revenue of $8000 is transferred from Emptying & Transport to the municipality. The annual net loss is the subtraction of treatment OPEX with revenues from end-use industry and emptying truck (emptying and transport). This financial flow analysis hints that emptying and transport, and treatment are being operated suboptimally. Using the current fleet of trucks and the estimated number of clients, 5900 emptying events was calculated, which averaged into 3 to 4 septic tanks emptied by each truck daily. Evaluating the truck capacity (4 and 6 m³), and assuming that each truck empty several septic tanks to its full capacity, this means that each truck only made one trip to the treatment facility a day. Assuming that the travel distance between 10-20 km, to make about 1 hour travel time maximum, each truck only operates from 4 to 5 hours a day, which considered to be half-day work. Subsequently, it was calculated that the treatment plant receives about 25 m³ FS daily, in which settle to digest in 31 tanks of the treatment plant. The retention time at the digestion tank is about 28 days (Taweesan *et al.* 2015). The tank dimensions were not reported, thus the treatment capacity could not be confirmed. Hence Author could not confirm if the treatment plant is under-utilised. (AIT 2015) .suggested that the treatment plant operated at 75% of design capacity.

The eSOS Monitor calculate both CAPEX and OPEX, but the CAPEX for the user interface and containment were not considered in these simulations, due to most toilets being incorporated into the structure of the home. Hence the CAPEX would be included in the rental or purchase price of the residence. Other CAPEX were included in the scenarios so full cost recovery can be evaluated for a series of scenarios (Table 9-5).

The eSOS Monitor calculates the financial flow in accordance to the material flow (e.g. generated FS) across the SVC. These cost functions requires various input data of unit prices of water, energy (electricity and fuel), technical data of the corresponding technical option (e.g. holding capacity of septic tanks, desludging trucks and treatment). For this elements, assumptions were made in the absence of actual data. Operational data was particularly difficult to obtain. Hence, the baseline scenario using data from Nonthaburi study case results in slightly different OPEX at each SVC components as a result of assumptions and rounding up the data. However, it was all verified with the existing data, that it does not differ more than 25% from the study case. Table 9-6 summarised the difference of data and results from baseline scenario.

Table 9-6 Comparison of Nonthaburi study case data and eSOS-Monitor's results

		User Interface and Storage ($)	Emptying and Transport ($)	Treatment ($)
CAPEX	Nonthaburi study case (AIT, 2015)	-	312,000	1,800,000
	Baseline scenario (eSOS Monitor)	-	312,000	1,798,000
OPEX	Nonthaburi study case (AIT, 2015)	-	59,000	115,000
	Baseline scenario (eSOS Monitor)	2,452,800	61,805	120,449
Revenues	Nonthaburi study case (AIT, 2015)	-	67,000	9,000
	Baseline scenario (eSOS Monitor)	-	83,827	9,000

In the first simulated scenario (Scenario 1), the emptying free was increased from $13 to $20 (over 50% increase – maximum reasonable increase) per toilet, to see if this would increase the financial sustainability of the emptying, transport and treatment (Table 9-5). It is acknowledged that this would require a change in the law as Public Health Act (1992) does not allow the public utility to charge above the current rate (Aecom & Eawag 2010). The extra financial burden on residents is relatively small, as OPEX for the user interface and containment rises by 1.6% (Table 9-7). The CAPEX is not covered by the revenue generated (Table 9-7) for the emptying, treatment and transport, although the increased in emptying fees increases the revenue for this part by 48% (Table 9-7). However, the system remains at financial loss. To cover for total OPEX alone, a significant increase in the emptying fee of more than $20 per toilet (i.e. approximately $30) would be required to make this system financially sustainable if FFM 2 is used. It is likely that this strategy would be accepted by the residents in the case study area.

Table 9-7 Results from the baseline scenario (data in Table 9-3) and Scenario 1 (data in Table 9-4), both using FFM 2

Part of SVC	Financial parameter	Baseline scenario ($)	Scenario 1 ($)
User interface and Storage	CAPEX	0	0
	OPEX	2,527,627	2,567,918
	Revenue	0	0
	EBITDA	-2,527,627	-2,567,918
	Net profit loss	-2,527,627	-2,567,918
	Payback	N/A	N/A
	BEP	N/A	N/A
	ROI	N/A	N/A
Emptying and Transport and Treatment	CAPEX	2,110,000	2,110,000
	OPEX	182,303	182,303
	Revenue	83,827	124,118
	EBITDA	-98,476	-58,185
	Net profit loss	-344,634	-304,352
	Payback	N/A	N/A
	BEP	N/A	N/A
	ROI	N/A	N/A

The results generated for the three other scenarios can be seen in Table 9-8. As the CAPEX for 'User Interface and Storage' is not included, then the rest of financial indicators are not reported. In Scenario 2 the emptying fee is increased from the baseline of $13 to $15 per toilet, and a discharge fee of $5 per 1.5 m³ (equal to volume per septic tank) is introduced. The increase was set to minimum (only $2 or 15% increase), combined with additional $5 per toilet, to be the same increase of $7 as total increase in Scenario 1. This scenario was run to provide alternatives should the emptying fee could not be increased to as high as in Scenario 1. This increased the OPEX for the user interface and containment by 0.5% (Table 9-7 and Table 9-8). The OPEX for the emptying and transport rose by $28,779 due to the introduction of a discharge fee, while the revenue increased by $11,512. It can be seen in Table 9-8 that the revenue nearly covers the OPEX for this part of the SVC. Although the treatment revenue increased from $9,000 to $37,780, this only covers approximately a quarter of the OPEX. Hence further budget support is required for this scenario and FFM to be financially sustainable.

Scenario 3 takes a multipronged approach to obtaining financial sustainability and introduces an annual sanitation tax of $50 (Table 9-5). The tax was set to $50 per household per year to get steady funding of $800000 from all households (16,000 households) being serviced by the system, not only households who empty their FS storage in the year. This increased the OPEX for the user interface and storage by 32%, which would be probably unpopular, but it should be noted that this is $4.16 per month and equates to 1.5% of the minimum wage (of $275) in Thailand. As the financial flows for emptying and transport remain the same as in Scenario 2, it can be seen that this portion of the SVC remains financially unsustainable. The sanitation tax is used as budget support for treatment, which means this part of the SVC becomes finically sustainable for the first time, meaning both the CAPEX and OPEX are recovered (Table 9-8) with a payback time of 29 months. In Scenario 4 the discharge fee was removed, and a discharge licence was introduced (Table 9-5). This led to a significant reduction in the OPEX for emptying and transport hence this part of the SVC becomes financially sustainable for the first time, with a payback period of approximately 13 years (Table 9-8). Although the removal of the discharge fee (as introduced in Scenario 2) reduced the revenue for treatment (Table 9-8) this part of the SVC remains financially sustainable, due to the budget subsidy provided by the sanitation tax.

Table 9-8 Results from Scenario 2 to 4 data in Table 9-4

Part of SVC	Financial parameter	Scenario 2 ($)	Scenario 3 ($)	Scenario 4 ($)
User Interface and Storage	OPEX	2,539,139	$3,339,193	$3,339,193
Emptying and Transport	CAPEX	321,000	321,000	321,000
	OPEX	90,584	90,584	62,005
	Revenue	86,339	86,339	86,339
	EBITDA	-4,246	-4,246	24,2334
	Net profit loss	-51,046	-51,046	-22,466
	Payback	N/A	N/A	154
	BEP	N/A	N/A	14,029
	ROI	N/A	N/A	N/A
Treatment	CAPEX	1,798,000	1,798,000	1,798,000
	OPEX	120,499	110,628	120,499
	Revenue	37,780	866,501	809,200
	EBITDA	-82,719	755,874	688,701
	Net profit loss	-265,519	576,074	598,801
	Payback	N/A	29	31
	BEP	N/A	3,806	4,177
	ROI	N/A	32	29

From the scenarios explored (Table 9-7 and Table 9-8), it can be seen that Scenario 4 is the only one which is financially sustainable across the SVC from emptying to treatment. The main reason for this is the use of a sanitation tax which provides a budget subsidy for the treatment process. The results gained in Table 9-7 can be used to explore further scenarios, i.e. from the results, it can be estimated that the if the sanitation tax in halved to $25 per household per year, that the payback time for treatment would be approximately five years, which is still a reasonable period of time using Model 3 or 4. Another strategy might be to add budget support raised via the sanitation tax to the emptying and treatment portion SVC in Models 3 or 4. This demonstrates the potential of using eSOS Monitor to optimise financial flows across the SVC for this case study area.

This chapter did not explore changing the components or parameters in the system, but from the data it was seen that volumetrically the trucks are being underutilised. They are only collecting sludge once per day, hence possibly discharging at the treatment plant once a day (as previously discussed in this section). For this to be convincingly changed or improved more information is required on the context, logistics and current practices. Additionally, the treatment plant may be working under capacity, but the exact treatment capacity could not be confirmed. What is clear is that the financial implications of these types of changes could be modelled using eSOS Monitor. Real-time tracking approach is incorporated into eSOS Monitor would definitely improve the transport and logistics and decrease OPEX.

9.3.1 eSOS Monitor Evaluation

The eSOS Monitor requires a large amount of technical data as input, which is difficult to obtain from literature. Operational data was particularly difficult to get, because they were often either not accessible, documented or reported. In this study case, the missing data could either be re-calculated from other data and triangulated or estimated from the experience or used as given by default in the program. CAPEX were deemed to be quite high when compared with the general purchase and or construction prices in Thailand. But the original high values were used as they originated from credible data sources.

It should be noted that the current version of eSOS Monitor is a beta version and as such is not particularly user-friendly. The amount of data required is similar to the data requirements for other financial tools such as Technical and Financial Toolkit (Aecom & Eawag 2010) and SANIPLAN (Ross et al. 2016). A lot of data input process could have been simplified. Unlike spreadsheet such as Excel where the formula and referred cells can be checked, it was difficult to check the data flow in this tool. Therefore, when there were some errors in the data calculation, it was not easy to figure if it is the mistake of data input, or if there is an error in the programming.

eSOS Monitor summed the costs as annualised CAPEX and annualised OPEX. Total costs of sanitation are commonly reported in NPV (Net Present Values) or annualised costs per capita (PAS-India 2015) or households. These financial indicators are not commonly used to measure financial flow in the sanitation sector. However, it was discussed in the previous section that those indicators are most useful for business users assessing business viability in FSM. Nevertheless, currently as it stands, most of the chains are not generating profit values hence some indicators e.g. EBITDA, BEP, ROI, etc. became invalid are not applicable, or values are negative.

9.4 Conclusions

A sanitation financial flow simulator was developed named the eSOS Monitor. This tool simulates the financial flow throughout and within the SVC. Once all required data are inputted, the tool can demonstrate changes in the financial flow when costs and fees are modified. Data from Nonthaburi was used to sucessfully validate this tool.

The tool was then used to simulate four scenarios, with the aim of increasing the systems financial sustainability. Those scenarios work around modifying emptying fees and introducing feasible instruments, i.e. sanitation tax, discharge fee and discharge license. Applications of these financial instruments aim to help recover the cost of emptying-transport, as well as treatment.

This chapter successfully demonstrates how the eSOS Monitor can be used to explore different FFMs for a case study area (Nonthaburi City Municipality) and how it can be used to optimise the financial sustainability across an SVC. Introducing a sanitation tax of $50 (to be paid by each household profited by the service of this sanitation system) and using this as budget support for treatement would lead to financial sustainability, although to execute this a change in the law would be required, and it could prove unpopular with residents. Applications of other financial instruments i.e. discharge fee or discharge licensing are less useful for the type of business model in Nonthaburi.

 The eSOS Monitor requires further development to ensure that it is user-friendly and to guide data collection, it shows great potential as it enables the exploration of financial sustainability for sanitation projects and programs. It is envisaged that it would be of use to planners and implementers from multiple sectors in both humanitarian and development sector.

References

Aecom and Eawag (2010). *A Rapid Assessment of Septage Management in Asia: Policies and Practices in India, Indonesia, Malaysia, the Philippines, Sri Lanka, Thailand, and Vietnam*, USAID.

AIT (2015). Faecal Sludge Management Financial Flow (Nonthaburi Case Study). In.

Bocken N. M. P., Short S. W., Rana P. and Evans S. (2014). A literature and practice review to develop sustainable business model archetypes. *Journal of Cleaner Production* 65(Supplement C), 42-56.

Branfield H. and Still D. (2009). *User manual for the WHICHSAN Sanitation Decision Support System*, Water Research Commission South Africa, Comission WR.

Campos L. C., Jain V. and Schuetze M. (2012). Simulating Nutrient and Energy Fluxes in Non-networked Sanitation Systems. In: *FSM2 Conference*, FSM2 Conference, Durban, South Africa.

Cowling R., Peal A. and Mikhael G. (2013). *100% access by design: a financial analysis tool for urban sanitation*, Water & Sanitation for the Urban Poor, WSUP, United Kingdom.

Dodane P.-H., Mbéguéré M., Sow O. and Strande L. (2012). Capital and Operating Costs of Full-Scale Fecal Sludge Management and Wastewater Treatment Systems in Dakar, Senegal. *Environmental Science & Technology* **46**(7), 3705-11.

Harada H., Schoebitz L. and Strande L. (2015). SFD Report Nonthaburi, Thailand. In, SFD Promotion Initiative, www.sfd.susana.org.

OECD (2017). *Geographical Distribution of Financial Flows to Developing Countries 2017*. OECD Publishing.

PAS-India (2015). SANIPLAN - A City Sanitation Planning Model. In, India.

Rao K. C., Kvarnstrom E., Di Mario L. and Drechsel P. (2016). *Business models for fecal sludge management*, International Water Management Institute (IWMI).

Ross I., Scott R. E., Mujica A., White Z. and Smith M. D. (2016). *Fecal sludge management: diagnostics for service delivery in urban areas-tools and guidelines*, World Bank, WSP WBG-.

Steiner M., Montangero A., Koné D. and Strauss M. (2003). Towards More Sustainable Faecal Sludge Management Through Innovative Financing–Selected Money Flow Options. *Swiss Federal Institute for Environmental Science and Technology (EAWAG), Department of Water and Sanitation in Developing Countries (SANDEC)*. http://www. eawag. ch/publications_e/e_index. html (February 5, 2005).

Strande L., Ronteltap M. and Brdjanovic D. (2014a). *Faecal Sludge Management: Systems Approach for Implementation and Operation*. IWA Publishing, London.

Strande L., Ronteltrap M., Brdjanovic D., Bassan M. and Dodane P. H. (2014b). *Faecal Sludge Management Systems Approach for implememntation and operation*. IWA Publishing.

Taweesan A., Koottatep T. and Polprasert C. (2015). Effective faecal sludge management measures for on-site sanitation systems. *Journal of Water Sanitation and Hygiene for Development* **5**(3), 483.

Tilley E. and Dodane P.-H. (2014). Financial transfers and responsibility in faecal sludge management chains. *Faecal Sludge Management. Systems Approach for Implementation and Operation. 1st ed. London, UK: IWA Publishing*, 273-91.

Trémolet S., Rama M. and Organization W. H. (2012). Tracking national financial flows into sanitation, hygiene and drinking-water.

UN-Global-Compact and KPMG-International (2015). SDG Industry Matrix: Financial Services. In.

WASHCost (2012). Providing a basic level of water and sanitation services that last: COST BENCHMARKS. In: *WASHCost Infosheet*, IRC, Netherlands.

Zakaria F., Garcia H. A., Hooijmans C. M. and Brdjanovic D. (2015). Decision support system for the provision of emergency sanitation. *Science of The Total Environment* **512-513**, 645-58.

10

Reflections and outlook

10.1 Reflections

The cholera outbreaks affecting displaced community in the aftermath of Haiti earthquake in 2010 brought humanitarian agencies' attention to inadequacy of emergency sanitation responses when the crisis occurred in urban settings. Initially the challenges were attributed by the difficulties to excavate pits as commonly practiced in many emergency scenarios. These challenges were then addressed by the deployment of raised latrines and the use of mobile latrines ranges from Peepoo bags, bucket latrines to Portaloos (chemical latrines). However, it was soon realised that these choices of latrines require collection or emptying services to sustain operation.

Identifying the problems, there are to three major issues to address. The first is that the sanitation provision needs to be thought to beyond toilet provisions, meaning that the pit emptying and other necessary services onwards to safe disposal of faecal matter should also be included in the emergency response plan. The second issue is linked to the first, that there is not sufficient technology options when it is not possible to dig a pit due to lack of space or difficult soil conditions. Emergency sanitation response has been pre-customised to choose for on-site sanitation system, and is limited to toilet or latrine provision only. An array of basic latrines, from communal model e.g. dedicated open defecation field and trench latrines, to cubicles of pit latrines are often the only thinkable technological choices. When these common options are not available, such as in the case of urban emergency in Haiti 2010, there was great confusions amongst the relief agencies. The use of chemical toilets was learned to be very costly due to importation costs and frequent servicing, and the application of peepoo bags and then raised latrines were not entirely successful, attributed to difficult collection or emptying and transport operations that were challenged by the crowded traffic, difficult access, and difficulties to find relatively safe disposal location.

The third issue to address is the lack of technical knowledge to plan for most suitable sanitation system. Most experiences were usually kept within relief agencies, while local authorities who are usually the first emergency responder do not have. Lack of technical knowledge and time constrain would likely to results in ineffective planning for emergency sanitation response. As it was learned from Haiti's experience, the portaloos solution was selected without anticipating the costly importation tax and intensive maintenance efforts.

More research and developments (R&D) to results in more rigorous sanitation responses were perceived to be in great needs following the challenges in urban emergencies. Relief providers' knowledge and field experiences are useful input, but they are less likely to have the capacity and resources to do thorough R&D, while doing their day-to-day jobs providing emergency reliefs. Being one of the few academic researchers in the topic of emergency sanitation, it gave Author the opportunity to carefully examine and rigorously think about improving the current emergency sanitation response delivery.

The thinking of this PhD research is centred to address identified issues i.e. provision beyond toilets/latrines i.e. emptying-transport-treatment-disposal operation, lack of suitable technology choices, and lack of technical knowledge to plan for effective emergency sanitation response. The first and second issue were addressed by developing a concept of integrated sanitation system, where operation of each sanitation chain is monitored. The concept is named emergency Sanitation Operation System, shortened into an acronym – eSOS (discussed in Chapter 2).

eSOS utilizes readily available technologies, particularly from information-communication technologies (ICT) established a concept to have an integrated sanitation system with ICT advancements at every chain. Following the several process entwined in the sanitation system, eSOS concept divides its proposed system into several infrastructures i.e. smart toilet, coordinated smart de-sludging unit and centralized treatment. These component are all monitored and coordinated using a monitoring eSOS software named eSOS Monitor. This PhD research have focused on parts of eSOS i.e. eSOS Smart Toilet and the governing software eSOS Monitor.

The design of eSOS Smart Toilet supported with eSOS Monitor aim to have a toilet system that can be constructed quickly, without ground excavation, and most importantly, can cope with high number and fluctuative usages. The eSOS Monitor is the governing software which tasks is to ensure responsive operation and maintenance to ensure continues toilet operations. Combined, eSOS Smart Toilet and eSOS Monitor became a new technology and operation management option for emergency sanitation solution. The development of experimental prototype of eSOS Smart Toilet and eSOS Monitor was discussed in Chapter 3.

The third issue of lack of technical knowledge to plan for appropriate emergency sanitation solution was addressed by the development of a decision support system (DSS), as discussed in Chapter 8. The DSS compiled appropriate technology solutions, organised in a logical selection mechanism to suit given emergency scenario.

Having a responsive operation and maintenance for emergency toilets would ensure continues operation of the toilets, as well as guaranty safe disposal of the faecal matter. Similarly, having a decision support tool to would assist in planning for appropriate emergency sanitation response in timely manner, even if the relief provider has limited technical knowledge. This research contributes to the emergency sanitation as a sector in providing a technological and managerial alternative in an urban emergency scenario through eSOS concept and eSOS Smart Toilet, as well as aiding preliminary decision making of emergency sanitation planning through the development of a DSS.

The research findings are summarised in the following sections, divided into 2 sections. The first section discusses findings from development and field testing of experimental prototype of eSOS Smart Toilet. The second section discusses DSS development and eSOS Monitor expansion.

10.1.1 eSOS Smart Toilet

This PhD research also supported the development of a smart toilet prototype that aims to ensure that the excrement containments are safely emptied before they are full. In addition, the smart toilet was also equipped with features that ease the toilet users. Amongst the conveniences are a built-in water supply system, a surface disinfection using UV lamp, careful design of space and dimension of the cubicle, smart lock system, alarm system, and lightings using light intensity sensors. The toilet is self-sufficiently powered by solar energy, captured by a solar panel on the roof top. Details were discussed in Chapter 3. All these features were put in place to support not only users, but also parties that operate and maintain the toilet. Toilet operator could always monitor the state of the toilet through eSOS Monitor.

To ensure that the toilet suffices its designed purposes and that is suitable for use in an emergency context, the toilet prototype was tested in a temporary settlement of a typhoon affected community in Tacloban City in The Philippines. The field test successfully gained insights of actual toilet usages. Different aspect of field test results were discussed from Chapter 4 to 7, and summarised as follows.

10.1.1.1   Use and operation

The eSOS monitoring software provides novel data[11] on toilet use and operations. The software makes use of data generated by weight sensors supported by manual data collection and door key provision generated information such as the duration of a visit, the amount of faecal sludge and urine generated per visit and per female/male or adult/child users. These data showed (see Chapter 4), among others, who the users were, whether the toilet equally appealed to male, female and child users, whether the toilet was also used at night time, if the toilet was only used for urination or that users were sufficiently comfortable to defecate. Manual observation alone would not have been able to answer all these questions, and continuous surveillance, such as by camera surveillance would have compromised the user's privacy.

Besides reporting usage data, continuous monitoring of the containment tanks ensured non-stop operation of the eSOS toilet, alerting the operator whenever the waste collection tanks require emptying.

The field test justified the service capacity of the eSOS toilet at its current design stage. Provided an emptying service is available whenever needed, the eSOS toilet was calculated to be able to accommodate over 200 visits a day.

10.1.1.2   Faecal sludge, urine, greywater and water supply characteristics

Münch et al. (2007) and Nyoka et al. (2017) highlighted necessary adjustments for users who performed cleansing with water in UD toilets in emergency camps. More than culture suitability, it also facilitates the faecal sludge (FS) treatment option. The eSOS Smart Toilet

---

[11] Detailed data about the toilet usage that has not been obtained by other studies to date

design which featured universal application in communities with different after-toilet cleaning cultures was evaluated in a disaster affected population who perform cleansing with water. Chapter 5 describes the characterisation of the faecal sludge, urine and grey-water produced by the test toilet. Moreover, as the toilet has a water supply and water treatment system, the characteristics of fed water and treated water were included in the assessment to measure water treatment effectiveness. The water treatment unit was able to treat the fed water to meet the Philippines' standard for recreational water.

FS from the test toilet was similar to fresh faecal sludge of medium to high-strength from a public toilet with a high number of visitors and frequent desludging. Subsequently, a high *E. coli* concentration was measured. The consistency was much wetter than FS collected from UD dry toilets. Therefore, instead of co-composting, anaerobic digestion was recommended as treatment. However, if the time and pathogen level are a concern, for example to avoid an epidemic outbreak, then thermal or chemical treatment was recommended as the safest and fastest treatment.

The urine and grey water samples from the test toilet contained *E. coli* as well. The *E. coli* concentration in the urine samples was high, requiring treatment before disposal to water bodies or use as fertiliser. Storage retention alone would not be sufficient and would take considerable time. The *E.coli* concentration in the grey water was low, and can be best treated via quick disinfection such as chlorination or thermal treatment, to meet discharge quality standard.

10.1.1.3   Effectiveness of features: UV-C light

Chapter 6 discusses the effectiveness of UV-C light installed in the eSOS Smart Toilet, which was evaluated in the laboratory in preparation for evaluation under real usage during the field research. In this case, UV-C light installed in the test toilet was able to partly inactivate *E. coli* during laboratory testing. However, inactivation of *E. coli* by the UV-C lamp could not be confirmed during the field test due to the absence of *E. coli* at sampling points. Nevertheless, the UV-C light inactivated total coliforms frequently detected in the samples. Hence, UV-C light has potential for toilet surface disinfection since total coliform is reported to be more resistant to UV than *E. coli*.

10.1.1.4   User Acceptance

Chapter 7 describes the response of the community to the test toilet. The in-built water supply system that includes an automatic water button and shower head were appreciated the most, due to the water scarcity problem at the test site and the community's custom to wash after toilet use with water. From the users' point of view, the ease of use and comfort of the toilet was more important than the potential health and environmental benefit.

The peculiarity of the toilet being high-tech did not prevent use. Nevertheless, it initially raised questions from residents at the test-site. The questions were fortunately voiced curiosity rather than concerns and emphasised the need for continuous communication between the sanitation provider and potential toilet users from the beginning.

For the user acceptance, the functionalities preceded other considerations (excluding economic factors, as the service was provided free of charge) to gain the users' acceptance of the smart toilet tested at their residing location. Furthermore, the toilet served the purpose from the research's viewpoint and the positive acceptance indicates a suitable application in emergency settings.

In summary, the smart features of eSOS toilet demonstrated merits during the field test, with some features having been tested in a laboratory setting prior to being dispatched to the Philippines, for example the effectiveness of the UV-C lamp to remove coliform bacteria. The field test provided feedback that user-operated smart features such as a water button that enables water to flow from the toilets in-built water system was most appreciated by the users, according to the user's acceptance study conducted during the toilet's field test. Most importantly, the advanced monitoring system of the eSOS Smart Toilet demonstrated expediency in achieving a responsive maintenance effort, which results in minimum service loss, in addition to optimum operation and maintenance costs.

10.1.2 Decision support tools for emergency sanitation and beyond

Timing has always been crucial in emergencies that leaves little room for a planning process. Planning on emergency sanitation response is no exception. Sanitation solutions are normally based on standard practice by relief agencies, without time available to opt for the best technology depending on the situation for a specific emergency. A decision support system (DSS) to assist making effective choices in sanitation planning in the realm of a crisis is necessary. A decision support system was developed during the PhD research to assist in choosing the most suitable sanitation technical option for a given emergency (See Chapter 8).

Studying the sanitation technical options that have previously been applied in emergencies, it was apparent that the sanitation solution was primarily of faecal sludge containment - on-site sanitation systems, without planning for the entire sanitation service chain. As a first step in the DSS development, technical sanitation options which have been used in emergencies or ones which have potentials to be used in emergencies, were inventoried. Gathered technology options were classified into function group in sanitation chain i.e. 'user-interface', 'collection/storage', 'conveyance' (emptying and transport), 'semi-centralised treatment 1', semi-centralised treatment 2, and 'disposal and re-use'. As a result of common emergency sanitation responses, where the responses focus on toilet provision only, there were a lot of technological options for 'collection/storage', but a limited number for the rest of the sanitation chain. A large part of the technical options list for other function groups had to be drawn from studying feasible technologies, some of them were still in the pilot or experimental stage. It was also observed that the list of technical options are expectedly predominantly of on-site sanitation systems, rather than sanitation systems comprising sewers. Additionally, the fact that some on-site sanitation systems are designed so that they do not require emptying/transportation or treatment, was to be accepted. Thus, the inclusion of 'no' options i.e. 'no-emptying/transport' and 'no-treatment'.

The second step was checking the compatibility of technologies in each function group, followed by an evaluation of the identified sanitation systems. Three evaluation criteria were used i.e. 'deployability', 'sustainability', and 'economic and environmental benefit'. The users are able to do the final evaluation by rating the selected technology options.

The DSS is a sanitation technology planning tool. Considerations regarding cost were only incorporated as user input in the evaluation part of the DSS; thus, it does not calculate the costs of the selected technology.

Recognising that financing sanitation is the key aspect to a sustainable sanitation system, it then became apparent that a decision support tool that simulates the financial flow of a sanitation system was necessary. Additional data from the development of the eSOS Smart Toilet that includes measurement of faecal sludge (FS) flows, in line with the eSOS concept that attempts to embed artificial intelligence in monitoring software, supported by the possibility of tracking devices at FS emptying vehicles, provides the opportunity to model the financial flow of a sanitation system. Consequently, a financial flow model - FFM (initially referred to as the eSOS business model, later named eSOS Monitor, validated in Chapter 9) was developed.

The eSOS Monitor was started as a governing software to monitor operational of eSOS Smart Toilet (see Chapter 3 and when it is being tested in Philippines in Chapter 4). The software was then expanded to function as a monitoring software not only for the toilet, but also to track the emptying and transport operation. Having all this information on faecal sludge flow enabled the eSOS Monitor to calculate the financial flow across the sanitation value chain. The software was then developed to its full potential (provided the data it gained from monitoring process), into a financial flow simulator. Hence it is able to advise policy maker and potential investors to a calculated financial flow based on the given sanitation value chain operation. It is able to provide information on costs and revenues from each chain, as well as provide a projection should any financial instrumentations such as taxation and subsidy schemes are introduced into the given operation scenarios. The validation of eSOS Monitor can be read in Chapter 9.

In order to validate the eSOS Monitor, a case study in Nanthaburi Municipality near Bangkok, Thailand was selected. The selection was principally because of the sufficient amount of projected data. The validation resulted in the close calculation of costs for emptying and operation, with some discrepancies from different assumptions and rounding up of data.

The simulator provides an overview of how an input from one sanitation chain is interlinked with another sanitation chain. It also shows cost and revenues from each sanitation chain, as well as opportunities for subsidy and tax schemes. Moreover, the simulator can simulate the changes when the fees or taxes are modified, suggesting whether the sanitation system is financially sustainable or not.

At this stage, eSOS Monitor can run independently from eSOS Smart Toilet. The eSOS Monitor incorporates all technology options in its simulations, including the eSOS Smart Toilet as one technology option.

10.2 Outlook

Since Haiti emergency in 2010, there has been a new realization of the importance of research and development (R&D) to find solutions for emergency reliefs, including in the WASH sector. Ever since, there has been also a shift from focusing primarily on water supply towards encouraging greater innovation in sanitation (Rush & Marshall 2015). Understandably, earlier established relief agencies such as Oxfam, Medecins Sans Frontier (MSF) and others intensified their R&D efforts (Rush & Marshall 2015). R&D efforts were also encouraged and occasionally endorsed by WASH Cluster, led by UNCEF. Academic institutions have been engaged to conduct researches on emergency sanitation, preferably in cooperation with the relief agencies. Additionally, developments from outside of humanitarian sector also play a role in facilitating R&D in emergency sanitation, such Re-invent the Toilet Challenge in 2012 that was initiated and funded by the Bill & Melinda Gates Foundation.

Availability of funding is also another driver of the changing landscape of emergency sanitation R&D. Certain collaborative such as Humanitarian Innovation Fund (HIF) (established in 2010) funded by DFID (UK's Department for International Development) has been having several WASH R&D initiatives. Their schemes has been funding relief agencies as well as academic institutions.

The sector continues to be more increasingly enthusiastic about R&D and innovations. In August 2016, Author contributed to the call to initiate an emergency sanitation compendium, seeing that it could use the knowledge from our emergency sanitation DSS (Zakaria *et al.* 2015). The compendium was published earlier 2018 (Gensch *et al.* 2018) with acknowledgement.

At the beginning, the concept of a smart toilet may have appeared to be too futuristic, but it has been widely accepted along with appreciation of novel technologies and innovations from other sectors. This has also been observed during the development of eSOS Smart Toilet experimental prototype. Soon after the eSOS Smart Toilet as a concept and toilet design became publically (3rd IWA Development Congress, Brdjanovic *et al.* (2013)), and later as news coverage of the prototype at Reuters (Reuters 2014) and Voice of America (VOA) (Hoke 2014), other examples of smart toilets appeared on the market with similar features to those of eSOS Smart Toilet e.g. eToilet in India (Pareek 2014). However, evaluation of other smart toilets has not been reported in the public domain yet. Despite of lack of documentations, emerging application of technological advances in toilets' design justifies demands and suitability of eSOS concept, not only in emergency, but also applicable in other, non-conventional urban settings. Images of the latest eSOS Smart Toilet design (eSOS Nairobi Toilet) and the e-Toilet from India are shown in Figures 10-1(a) and (b).

Observing the R&D initiatives by academic institutions, there are also growing interests in technology applications in sanitation systems. There has been emerging innovations in sanitation using sensors and monitoring/controlling features, to mention a number of innovations, ranging from proposed application of sensors at the toilet's containment tank in railway coaches in India (Chintan *et al.* 2015; Kadge *et al.* 2016), water saving feature in urinal (Osathanunkul *et al.* 2017), and another smart toilet concept, also in India (Elakiya *et al.* 2018). In addition there has been a research applying sensor to monitor household latrine usages (O'Reilly *et al.* 2015) and GPS trackers on desludging trucks for monitoring purpose (Schoebitz *et al.* 2017).

The absolute explosion of different options for toilets and on- and off-site treatment has been the result of the Reinvent Toilet Challenge, with two fairs (Seattle 2012 and Chennai 2014). Videos about the innovations exhibited in both fairs are available on youtube (GatesFoundation 2012; SuSanA 2014)

This PhD research succeeded in using technology advances to develop a technology and tools to assist in optimizing the efficiency of the emergency sanitation operation to meet public health objectives in providing sanitation to disaster affected community. These advances have been applied to a smart toilet design; also to decision support tools as demonstrated in this study by the development of decision support system to select most appropriate sanitation system, which then later extended into a financial flow simulator i.e. eSOS Monitor.

10.2.1 Ways forward

Challenges remain to be addressed both in research and non-research related aspects. Some of these issues are discussed as follows. The experimental prototype of eSOS Smart Toilet was designed to serve mainly as testing platform. Therefore this test toilet has been assembled with the full spectre of functionalities that includes various sensors, water supply with built-in water treatment unit, UV-C lamp, and many others. However, the prototype design was not having a modular structure as intended for the final eSOS Smart Toilet product. Another design features yet to be applied in the test toilet is the integrated hollow walls that function as water and greywater containments. Instead, the test toilet was equipped with tanks at the exterior, each for water supply and collection of grey water.

In addition, several features of the experimental toilet have been introduced to provide high accuracy data, information and measurements for research purposes. The new generation of practical prototype will likely be equipped with less precise sensors, to reduce costs and fit the new purpose.

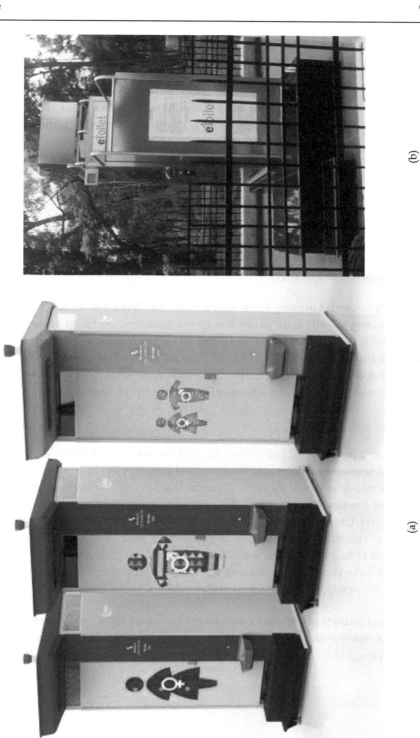

(a)

(b)

Figure 10-1 (a) the latest design of eSOS Smart Toilet; (b) The e-Toilet invented by a company from India, appeared in the market in September 2014 (https://www.thebetterindia.com/13940/etoilet-changing-way-public-sanitation-works-india-eram-marico/

Evaluating application of smart toilets in emergencies and how humanitarian agencies would welcome such idea, we have made some observations when presenting a paper on eSOS smart toilet development in a humanitarian innovation conference in Oxford, UK, in 2014, which was largely attended by relief and funding agencies. The responses were mainly encouraging. However, there were concerns about costs, as well as the impressions that the many smart features would be overwhelming for an emergency situations that requires moderations to.

At the moment, an improved design of eSOS Smart Toilet that took into consideration the findings from the field test is being manufactured. After successful testing of the eSOS experimental toilet in transitional settlement in Tacloban City, Philippines (2015, Technology Readiness Level – TRL 5), agreement with sponsors (i.e. Dutch government viaVia Water) has been reached to bring the eSOS Smart Toilet development to RTL 7 that will make the product closer to the ultimate goal of RTL 9 and commercial exploitation. This project is called eSOS Nairobi Toilet (https://www.viawater.nl/projects/esos-smart-toilet-kenya).

The latest version of eSOS Monitor still needs to be tested in harsh condition in peri urban areas of Nairobi, and to be further validated with different study cases to test its versatility and on different scenarios. eSOS Monitor has successfully simulated financial flow across the sanitation value chain, with interconnection between chains being the biggest achievement. But whilst the simulators is now available, gathering input data remains a challenge.

***

Finally, although the use of technological advancements in sanitation was initially considered as overly futuristic, it has become rather necessary than the luxury in the increasingly complex and challenging world. Moreover, as for emergency sanitation, there is a factual trend for more emergencies occurring and to occur in urban areas, where it becomes a necessity to have on-site sanitation system with onward management to safe disposal. Further, more extreme scenarios such as the case of Ebola emergency in western and central Africa, requires much improved sanitation management to prevent the disease outbreak. Advanced technology supported interventions (such as smart toilets), as well effective decision making aided by decision support tools that ensures safe handling of faecal sludge would be undoubtedly useful in these situations.

In addition, an innovation goes through stages of development before the invention can be used as a product, starting from the gaps analysis, ideas, design, development, field testing, diffusion, adoption, and finally scaling up. This PhD research tried to actualise the front end until mid of these innovation stages, but it has not reached (neither it was planned to do so) the ultimate milestone of the process. Academic institutions with their research capacity are crucial in developing ideas and experimental prototype, but the industry is essential to pick up and enhance the development, prototyping and commercialisation. As it was experienced in this research, the development of the eSOS Smart Toilet was possible by cooperation between an academic institution and a design company and software company. During the field test, the research team obtained help and support from a humanitarian organisation (i.e. Samaritan Purse), with cooperation and willingness from the community to participate at the test site (i.e. Abucay Bunkhouse community). This all confirms that a collaborative and multi-disciplinary efforts is needed to bring an idea to working invention. Unfortunately, it still takes too long (up to 10 years) to complete such process.

Emergency sanitation (as a sector) urgently needs more applied innovations to be able to provide more effective responses. The sector of emergency sanitation gave the impression of exclusivity for certain stakeholders, with relief agencies claiming that it is their domain. Only few academic or research institutes have 'privilege' so far to be considered as research partners when the relief agencies requires some scientific testing of certain innovations. Despite their increased R&D capacities, relief agencies needs to further strengthen links with academic institutions and relevant industries. Emergency sanitation dominated by practitioners, should become more open to changes and realize innovations needs time. Lessons learned and success stories need to be shared, and combined strength in joining forces between relief agencies, academic institutions and relevant industries will surely accelerate filling in the innovation gaps in emergency sanitation.

References

Brdjanovic D., Zakaria F., Mawioo P. M., Thye Y. P., Garcia H., Hooijmans C. M. and Setiadi T. (2013). eSOS® innovative emergency sanitation concept. In: *3rd IWA Development Congress and Exhibition*, Nairobi, Kenya.

Chintan P., Jatin Y., Sanket K. and Adeshara D. (2015). Autometic working bio-toilet tank for railway coaches. *International Journal of Advance Engineering and Research Development* 2(10).

Elakiya E., Elavarasi K. and Priya K. (2018). Implementation of smart toilet (swachh shithouse) using iot embedded sensor devices. *International Journal of Technical Innovation in Modern Engineering & Science* 4(4).

GatesFoundation (2012). Reinvent the Toilet Fair - Highlights | Bill & Melinda Gates Foundation. In.

Gensch R., Jennings A., Renggli S. and Reymond P. (2018). Compendium of Sanitation Technologies in Emergencies. In. 1st edn, German WASH Network, Eawag, WASH Cluster, SuSanA.

Hoke Z. (2014). New Smart Toilet to Prevent Water Pollution in Disaster Areas. https://www.voanews.com/a/new-smart-toilet-to-prevent-water-pollution-disaster-areas/1971723.html (accessed 15 August 2018).

Kadge A., Varute A., Patil P. and Belukhi P. (2016). Automatic Sewage Disposal System for Train. *International Journal of Emerging Research in Management &Technology* 5(5), 87-91.

Münch E. V., Ochs A., Amy G., Mwase H. and Fesselet J. F. (2007). Provision of sustainable sanitation in emergency situations: Role of ecosan. In: *Sustainable Development of Water Resources, Water Supply and Environmental Sanitation: Proceedings of the 32nd WEDC International Conference*, pp. 506-9.

Nyoka R., Foote A. D., Woods E., Lokey H., O'Reilly C. E., Magumba F., Okello P., Mintz E. D., Marano N. and Morris J. F. (2017). Sanitation practices and perceptions in Kakuma refugee camp, Kenya: Comparing the status quo with a novel service-based approach. *PLOS ONE* **12**(7), e0180864.

O'Reilly K., Louis E., Thomas E. and Sinha A. (2015). Combining sensor monitoring and ethnography to evaluate household latrine usage in rural India. *Journal of Water Sanitation and Hygiene for Development* **5**(3), 426-38.

Osathanunkul K., Hantrakul K., Pramokchon P., Khoenkaw P. and Tantitharanukul N. (2017). Configurable automatic smart urinal flusher based on MQTT protocol. In: *2017 International Conference on Digital Arts, Media and Technology (ICDAMT)*, pp. 58-61.

Pareek S. (2014). This eToilet Is Changing The Way Public Sanitation Works In India. https://www.thebetterindia.com/13940/etoilet-changing-way-public-sanitation-works-india-eram-marico/ (accessed 15 August 2018 2018).

Reuters (2014). NETHERLANDS: The 'Smart' toilet which could replace hole-in-the-ground disaster zone sanitation. https://reuters.screenocean.com/record/344570 (accessed 15 August 2018 2018).

Rush H. and Marshall N. (2015). *Case study: Innovation in Water and Sanitation and Hygiene*, CENTRIM, University of Brighton, DFID, United Kingdoms.

Schoebitz L., Bischoff F., Lohri C. R., Niwagaba C. B., Siber R. and Strande L. (2017). GIS analysis and optimisation of faecal sludge logistics at city-wide scale in Kampala, Uganda. *Sustainability* **9**(2), 194.

SuSanA (2014). Reinvent the Toilet Fair, Delhi, India (overview tour of the exhibits). In.

Zakaria F., Garcia H. A., Hooijmans C. M. and Brdjanovic D. (2015). Decision support system for the provision of emergency sanitation. *Science of The Total Environment* **512-513**, 645-58.

About the Author

Fiona Zakaria was born in Bandung, Indonesia. Indonesia is a country constantly experiencing onsets of natural disasters ranged from most common natural disaster such as floods and typhoons, to less common one such as volcanic eruptions, even least common such tsunamis. In addition, Fiona is an Achenese and she grew up in Banda Aceh. 'Acheh' or locally written as 'Aceh' is an ethnic group in Indonesia living in Aceh Province. This province was torn apart by over 30-years of an armed conflict between Indonesian government and a separatist rebel group. And so, crises and emergencies are not unfamiliar to Fiona.

As soon as finishing a bachelor degree in civil engineering from Universitas Brawijaya in East Java, Indonesia, Fiona worked for International Committee of The Red Cross (ICRC) as a field engineer in Water and Habitat division, working to provide humanitarian response (water and sanitation) to conflict affected communities in Aceh. That was during the peak of armed conflict in Aceh, until it was resolved in 2005, immediately after the 30-meter tsunami tidal wave hit the west coast of Aceh in December 2004.

The 2004-tsunami also sent 'humanitarian tsunami' waves to Aceh. More than 200 different humanitarian agencies responded. It was, still to date, one of the biggest humanitarian operations in the world. Fiona volunteered for a short period for ICRC in the first month in the tsunami aftermath. She too, as many other Acheneses, had lost family members in the disaster. But shortly after the volunteering work, she left to finish her MSc study in Sanitary and Environmental Engineering, at Universiti Putra Malaysia.

After finishing this MSc degree mid-2006, Fiona returned to Aceh and soon was recruited as Water Environmental Sanitation (WES) Specialist for United Children's Funds (UNICEF). She was assigned for different Water Sanitation and Hygiene (WASH) programs implemented in East Coast, and later also West Coast of Aceh, as part of the recovery responses after tsunami and armed conflict. When this recovery programs was about to end in mid-2008, Fiona won an opportunity to do another master study in the United Kingdom by Chevening, a prestigious scholarship scheme by the Foreign Commonwealth Office. With this scholarship, she resigned from UNICEF to study Water Science, Policy and Management at University of Oxford, UK.

Immediately after obtaining MSc degree from Oxford, she was recruited as a WASH consultant, back at UNICEF, to work in South Darfur, Sudan. Fiona worked for the entire year of 2010 for this position, before she returned to Indonesia. However, the development in Sudan that underwent political turmoil to result in the referendum of South Sudan has resulted in funding re-allocating, and so the offer to return to work in Sudan was withdrawn. She spent the year of 2011 in Indonesia doing some short consultancy works, including one for WASH Cluster Indonesia (coordinated by UNICEF). Early 2012, after a long job-search, having applied and shortlisted to a number of positions both in the country and internationally – but not a single job offer, she was offered a position as a PhD researcher at (was) UNESCO-IHE, Delft, the Netherlands, in a topic that was close to her living and working experiences - 'emergency sanitation'.

Five years went on, Fiona was recruited as a post-doctoral researcher at School of Civil Engineering, University of Leeds, UK. She is doing a research in costings of urban sanitation applicable in developing countries.

List of Publications

JOURNAL ARTICLES

Zakaria F., Curko J., Muratbegovic A, Garcia H. A., Hooijmans C.M., and Brdjanovic D. (2018) Evaluation of a smart toilet in an emergency camp. *International Journal of Disaster Risk Reduction 27, 512 - 523*

Zakaria F., Thye Y.P., Hooijmans C. M., Garcia H. A. Spiegel A.D. and Brdjanovic D. (2017). Use acceptance of the eSOS™ Smart Toilet in a temporary settlement in the Philippines. *Water Practice & Technology 12(4), 832-847.*

Zakaria F., Harelimana B., Ćurko J., van de Vossenberg J., Garcia H. A., Hooijmans C. M. and Brdjanovic D. (2016). Effectiveness of UV-C light irradiation on disinfection of an eSOS™ smart toilet evaluated in a temporary settlement in the Philippines. *International Journal of Environmental Health Research 26(5-6), 536-53.*

Brdjanovic D., Zakaria F., Mawioo P. M., Garcia H. A., Hooijmans C. M., Ćurko J., Thye Y. P. and Setiadi T. (2015) eSOS™ – emergency Sanitation Operation System, *Journal of Water, Sanitation and Hygiene for Development* 5 (1), 156–164.

Zakaria F., Garcia H., Hooijmans C., and Brdjanovic D. (2015) Decision support system for the provision of emergency sanitation. *Science of The Total Environment 512, 645-658.*

Katayon, S., Fiona, Z., Noor, M.J. Megat Mohd, Halim, G. Abdul and Ahmad, J. (2008). Treatment of mild domestic wastewater using subsurface constructed wetlands in Malaysia. International Journal of Environmental Studies, 65 (1), 87-102.

Furlong C., Zakaria F., and Brdjanovic D. (2018) Development and Validation of a financial flow simulator for the sanitation value chain. *In preparation*

CONFERENCE PRESENTATIONS

Zakaria F., Curko J., Muratbegovic A, Garcia H. A., Hooijmans C.M., and Brdjanovic D. (2018) Evaluation of a smart toilet in an emergency camp. In *UNC Water and Health Conference – Where science meets policy*, 28 October – 2 November 2018, Chapel Hill, United States

Zakaria F., Harelimana B., Ćurko J., van de Vossenberg J., Garcia H. A., Hooijmans C. M. and Brdjanovic D. (2015) Effectiveness of UV-C light irradiation on disinfection of an eSOS™ smart toilet evaluated in a temporary settlement in the Philippines. In *Environmental Technology and Management Conference*, 23-24 November 2015, Bandung, Indonesia

Zakaria F., Curko J., Muratbegovic A, Garcia H. A., Hooijmans C.M., and Brdjanovic D. (2014) eSOS™ smart toilet development. In *Conference on Humanitarian Innovation*, 19–20 July 2014, Oxford, United Kingdom

Zakaria F., Garcia H., Hooijmans C., and Brdjanovic D. (2013)Decision Support System for Emergency Sanitation Provision. In *Third IWA Development Congress*, 14 – 17 October 2013, Nairobi, Kenya

Acknowledgement

'Sanitation' as a general discussion topic has been largely anecdotal to a lot of people. I learned from own experiences, as well from other sanitation engineers, mentioning the word 'shit' is a turn-off in general discussions. Academia tried to tone the reputation of the word by using more dignified words such as 'faeces' or 'excreta', but still, sometimes the best way to bring the topic to the discussion table is by mentioning the word 'shit' and I had fun spotting persons that either gasp or giggle in the room.

This PhD research has been about shit management in emergencies, and I would like to thank the followings.

First, I'd like to thank my promotor Professor Damir Brdjanovic, for thinking about sanitation in emergencies and include it as one of research areas in his research project proposal to the Bill and Melinda Gates Foundation. I believed that there has been nobody else to offer a PhD fellowship in this line of topic. I also thank him for giving me the opportunity to do the PhD research. I was told that he selected me out of some two hundreds applicants. And that was only the beginning, Prof. Brdjanovic was a supervisor that oversee my entire research and put everything together for me, not just for one project deliverable, but also for the ambition to contribute to better sanitation response in emergencies. I have been enjoying our professional interactions, and I have been learning a lot from him.

I also would like to thank my mentors, Tineke and Hector to provide me with suggestions and advice all the way through. I would not have achieved what I have achieved without their kind help.

The development of the experimental eSOS Smart Toilet and then the field research was the big part of my PhD research. I thank everyone in the eSOS toilet team that has made it possible, Josip Curko from University of Zagreb, Ronald Lewerissa and Jacco de Haan from Flex/design, Ahmed Muratbegovic and his team/associates (Emir, Aldin and Edin) from SYSTECH Bosnia.

Outside of eSOS team, I need to thank people of IHE lab, Fred Kruis, Peter Heerings, Lyzette Robbemont and Berend Lolkema for their help when I was preparing to do the field research. Along writing-ups and doing research, I am thankful for helping hands from Jack de Vossenberg on microbiology, swabbing tests procedures, and related matters; Mariska Ronteltap as a resourceful person for urine and faecal sludge characterisation, treatment and reuse; Yoke Pean Thye – for the help during the field research, co-authoring an article and potential project associate (too bad we didn't get that HIP project, but we'll have other opportunity to work together, I am sure), Prof. Mugsy from University of Cape Town for advice on the user acceptance study, and Claire Furlong that helped me completing the chapter on financial flow model.

To the people of Abucay Bunkhouse who have welcomed the eSOS research team and helped us in anyways that they could. I am glad that they have now settled in new houses. My sincere hopes for their well being. To Samaritan Purse Philippines in Tacloban office, who have assisted to more than I could have expected.

In addition, my sincere gratitude to Mloelya Mwambu and Bertin Harelimana, master students whose MSc thesis were part of my research.

Outside of the research life, I am grateful to have friends around me in IHE. Some friends contributed to both research and social life. This is to mention Isnaeni Murdi Hartanto that helped me putting all the VBA coding together for the DSS. Yuli Ekowati helped me with references to microbiology, and other kinds of help throughout my stay in Delft. I owe my enjoyable time in Delft to the 'sanitary gang', they were my office mates plus Nikola (who pushed his way as part of the gang although uninvited). The sanitary gang members are Laurens, Javier, Sondos, Mr. Kim, Peter and Joy. Sondos was the bravest with that email incident. Laurens defined what exotics and nots (which was totally unacceptable – he has different takes on everything anyway), Javier is a brother to me, and he has been the only source to learn beautiful Spanish names, Mr. Kim taught me about wisdom and patience, while Peter and Joy gave a taste of Kenya to the room. The incident with the office lamp cemented our close friendships onwards. I am sorry we have not got the chance to redo the BBQ on the secluded rocky part of Scheveningen. There is also my Southeast Asian alliance with food name code– Yuli (anything for free is most delicious), Linh (Pho, fresh lumpia, mussel and fish sauce), Shah (bakso, mee curry and roti jala), Anuar (food provided by his wife is always tasty), Isnaeni (sate gulung), Tarn (pad thai, panang, tom yam with remon glass), Pak Aries, Dikman, Leo, Soni, Mba Mawiti who is closely associated with any dining invitations, and of course Pak Suryadi for numerous invitations to eat at his place or at favourite restaurants of choices and also all the 'oleh-oleh' (I appreciate the kindness, nobody else care for us like you do, Pak).

Other IHE friends outside of these groups I need to thank is Nirajan (I can't mention all the 'weird statements' I have about you here) and Anita, Motasem and Hanaa (thank you for teaching me proper Qoran reading), and Mohanad (the only one offered to care for my orchids when I was away), also friends at Muslim Students Association, and members of PhD Association Board (PAB) 2012-2014. Over the years, there have been many PhD-ers and MSc-ers that I could not mention here one by one, who have left IHE or still around Delft, but they all have made my life in Delft enjoyable.

I was saved from the humiliation of the long PhD completion as I was recruited by Prof. Barbara Evans, a sanitation champion hailed from University of Leeds. She continued motivating me to complete this PhD. I also received helps and supports from my colleagues in Leeds, Sally, Dani, Celia, Gio, Tristano and Miller.

Finally to my mom, I know you were not in favour of this decision to pursue PhD at the beginning, and it was difficult to explain what research that I actually did (toilet, excreta, refugee camps, what??, why???), but I do appreciate that you let me do it anyway. To my sister and brother in law to take care of Mom and the moral supports, to my nieces Wafa and Haybah to release me from birthday-gifts obligations (we have a signed agreement for this). My father would have been the proudest of this achievement. This dissertation is dedicated in the loving memory of my father, Dr. Dipl. Ing. Zakaria Ismail.

SENSE

Netherlands Research School for the
Socio-Economic and Natural Sciences of the Environment

D I P L O M A

For specialised PhD training

The Netherlands Research School for the
Socio-Economic and Natural Sciences of the Environment
(SENSE) declares that

Fiona Zakaria

born on 2 April 1977 in Bandung, Indonesia

has successfully fulfilled all requirements of the
Educational Programme of SENSE.

Delft, 28 June 2019

the Chairman of the SENSE board

Prof. dr. Martin Wassen

the SENSE Director of Education

Dr. Ad van Dommelen

The SENSE Research School has been accredited by the Royal Netherlands Academy of Arts and Sciences (KNAW)

K O N I N K L I J K E   N E D E R L A N D S E
A K A D E M I E   V A N   W E T E N S C H A P P E N

The SENSE Research School declares that Ms Fiona Zakaria has successfully fulfilled all requirements of the Educational PhD Programme of SENSE with a work load of 39.3 EC, including the following activities:

SENSE PhD Courses

- Environmental research in context (2013)
- Research in context activity: "Preparing and creating two explanatory video presentations for the eSOS toilet: prototype development and field testing" (2014-2015)

Advanced MSc Courses

- Faecal sludge management, UNESCO-IHE (2013)

Management and Didactic Skills Training

- Deputy chair of the PhD Association Board, UNESCO-IHE (2012-2014)
- Moderating sessions at the PhD symposium, UNESCO-IHE (2012-2014)
- Supervising two MSc students with thesis entitled 'Development of a conceptual framework for decision support system for emergency sanitation' (2012-2013) and 'Analysis of UV disinfection and cleaning method from field testing of eSOS smart toilet in an emergency settlement' (2014-2015)
- Teaching in the summer course 'Humanitarian water, sanitation and hygiene' (2013)
- Teaching on 'Emergency sanitation competition' in Faecal Sludge Management Module 11, UNESCO-IHE (2013)
- Teaching on 'Development of the eSOS toilet' in Faecal Sludge Management Module 11, UNESCO-IHE (2014)

Selection of Oral Presentations

- *Decision support system for emergency sanitation provision.* 3rd International Water Association (IWA) Development Congress, 14-17 October 2013, Nairobi, Kenya.
- *Emergency WASH related research in UNESCO-IHE,* International Women's Day conference on 'Tapping into the unlocked potential of women in emergencies', 7 March 2014, Delft, the Netherlands.
- *eSOS® smart toilet development,* Humanitarian Innovation Conference (HIP2014), 19-20 July 2014, Oxford, United Kingdom
- *Effectiveness of UV-C light irradiation on disinfection of an eSOS® smart toilet evaluated in a temporary settlement in the Philippines.* Environmental Technology and Management Conference (ETMC2015), 23-24 November 2015, Bandung, Indonesia

SENSE Coordinator PhD Education

Dr. ir. Peter Vermeulen

T - #0008 - 170819 - C220 - 240/170/12 - PB - 9780367361815